PHYSICS RESEARCH AND TECHNOLOGY

AF773352

ACOUSTIC CAVITATION THEORY AND EQUIPMENT DESIGN PRINCIPLES FOR INDUSTRIAL APPLICATIONS OF HIGH-INTENSITY ULTRASOUND

PHYSICS RESEARCH AND TECHNOLOGY

Additional books in this series can be found on Nova's website under the Series tab.

Additional E-books in this series can be found on Nova's website under the E-book tab.

PHYSICS RESEARCH AND TECHNOLOGY

ACOUSTIC CAVITATION THEORY AND EQUIPMENT DESIGN PRINCIPLES FOR INDUSTRIAL APPLICATIONS OF HIGH-INTENSITY ULTRASOUND

ALEXEY S. PESHKOVSKY
AND
SERGEI L. PESHKOVSKY

Nova Science Publishers, Inc.
New York

For permission to use material from this book please contact us:
Telephone 631-231-7269; Fax 631-231-8175
Web Site: http://www.novapublishers.com

LIBRARY OF CONGRESS CATALOGING-IN-PUBLICATION DATA
Peshkovsky, Alexey S.
Acoustic cavitation theory and equipment design principles for industrial applications of high-intensity ultrasound / Alexey S. Peshkovsky and Sergei L. Peshkovsky.
p. cm.
Includes index.
ISBN 978-1-61761-093-6 (softcover)
1. Ultrasonic waves--Industrial applications. 2. Cavitation. 3. Ultrasonic equipment--Design and construction. I. Peshkovsky, Sergei L. II. Title.
TA367.P47 2010
620.2'8--dc22

2010029735

Published by Nova Science Publishers, Inc. † New York

CONTENTS

PREFACE

A multitude of useful physical and chemical processes promoted by ultrasonic cavitation have been described in laboratory studies. Industrial-scale implementation of high-intensity ultrasound has, however, been hindered by several technological limitations, making it difficult to directly scale up ultrasonic systems in order to transfer the results of the laboratory studies to the plant floor. High-capacity flow-through ultrasonic reactor systems required for commercial-scale processing of liquids can only be properly designed if all energy parameters of the cavitation region are correctly evaluated. Conditions which must be fulfilled to ensure effective and continuous operation of an ultrasonic reactor system are provided in this book, followed by a detailed description of "shockwave model of acoustic cavitation", which shows how ultrasonic energy is absorbed in the cavitation region, owing to the formation of a spherical micro-shock wave inside each vapor-gas bubble, and makes it possible to explain some newly discovered properties of acoustic cavitation that occur at extremely high intensities of ultrasound. After the theoretical background is laid out, fundamental practical aspects of industrial-scale ultrasonic equipment design are provided, specifically focusing on:

- electromechanical transducer selection principles;
- operation principles and calculation methodology of high-amplitude acoustic horns used for the generation of high-intensity acoustic cavitation in liquids;
- detailed theory of matching acoustic impedances of transducers and cavitating liquids in order to maximize the ultrasonic power transfer efficiency;

- calculation methodology of "barbell horns", which provide the impedance matching and can help achieving the transference of all available acoustic energy from transducers into liquids. These horns are key to industrial implementation of high-power ultrasound because they permit producing extremely high ultrasonic amplitudes, while the output horn diameters and the resulting liquid processing capacity remain very large;
- optimization of the reactor chamber geometry.

Chapter 1

INTRODUCTION

A multitude of important physical and chemical processes promoted by ultrasonic cavitation can be implemented on industrial scale by utilizing high-capacity flow-through ultrasonic reactor systems. These systems can permit processing large volumes of liquids and commonly comprise an ultrasonic-frequency electrical signal generator, an electromechanical transducer which converts the electrical signals into ultrasonic vibrations, an ultrasonic horn, which amplifies and transmits the vibrations into the liquids, and a flow-through reactor chamber (flow cell) which contains the flowing liquids. A general schematic of such a system is presented in Figure 1 [1, 2]. Several conditions must be fulfilled in order to ensure effective and continuous operation of an ultrasonic reactor system:

a. technologically necessary intensity of ultrasonic cavitation must be achieved in the liquid;
b. size and homogeneity of the cavitation region formed in the liquid must be maximized (well developed cavitation region);
c. reactor chamber must direct all of the liquid through the cavitation region (no liquid bypass);
d. electromechanical transducer must be electrically save, capable of continuous operation at full power for extended periods of time, and able to provide high radiation power levels;
e. ultrasonic horn must be capable of amplifying vibration amplitudes (high gain) while maintaining maximum possible size of the resulting cavitation region (large output diameter);

f. mechanical stresses present in the electromechanical transducer and the ultrasonic horn must not approach the limiting fatigue strength values for the corresponding materials;
g. entire system as well as each of its components must not be in danger of becoming overheated during continuous operation at full power.

High-quality engineering calculations of ultrasonic reactor system components can only be properly performed if all energy parameters of the cavitation region are correctly evaluated, since this region represents the active acoustic load of the electromechanical transducer (through the ultrasonic horn) and is the target "consumer" of all produced ultrasonic energy. We will, therefore, start by providing a detailed model of acoustic cavitation, explaining the mechanism by which the ultrasonic energy is absorbed in the cavitation region. A discussion of design principles of the main ultrasonic reactor system components will follow.

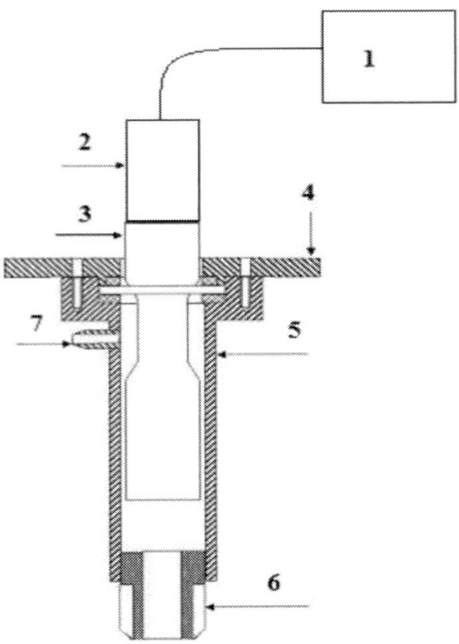

Figure 1. Schematic of an ultrasonic reactor system is presented. 1 – ultrasonic electrical generator, 2 – electromechanical transducer, 3 – ultrasonic horn (in this case, barbell horn), 4 – mounting flange, 5 – reactor chamber, 6 – working liquid inlet, 7 - working liquid outlet.

Chapter 2

SHOCK-WAVE MODEL OF ACOUSTIC CAVITATION

In order to properly design powerful ultrasonic sources for ultrasonic reactors, it is necessary to know the exact value of the intensity of acoustic energy radiated into the working liquid. This information is usually obtained experimentally because no adequate physical model of acoustic cavitation that would allow one to obtain such data through calculation exists. The development of an adequate model of acoustic cavitation, although of great importance, has in the past been severely restricted by considerable mathematical difficulties associated with the necessity of finding numerical solutions to nonlinear equations describing the cavitation region (visible region of large cavitation bubble population) [3]. The utilized direct analytical solutions of these equations in different approximations do not give practical results suitable for the design of ultrasonic equipment [4, 5].

The literature on acoustic cavitation mainly tends to involve numerical models of spatio-temporal characteristics of the cavitation region [6-8]. Large number of theoretical acoustic cavitation models has been developed along with the corresponding methods of numerical analysis of such models. Computer simulation-based investigations of acoustic cavitation have also been proposed, involving complex non-linear physicomathematical models and including many aspects of spatial movement of cavitation bubbles in an acoustic field, spatial distribution of the characteristics of these fields in a liquid, interaction between the bubbles themselves, properties of acoustical flow, etc [9-12]. Water is most frequently used for the experimental verification of such theoretical models.

No adequate explanation of the mechanism by which dissipation of the primary acoustic energy of a radiator occurs in a liquid at cavitation is, however, available from the literature. Additionally, no theoretical method permitting to calculate this energy in a manner adequate to the available experimental data currently exists. Meanwhile, the exact knowledge of the mechanisms by which heating of a liquid in the presence of cavitation-inducing acoustic waves occurs is important not only for the understanding of the related sonochemical processes, but also for the practical design calculations that would permit constructing improved high-capacity ultrasonic radiators and reactors.

2.1. Visual Observations of Acoustic Cavitation

Several authors provided common [13], high-speed [14] and stereoscopic high-speed [15] photographs of the cavitation region, obtained in the presence of relatively low-intensity acoustic fields. At these conditions, the cavitation region is located some distance away from the radiating surface and has a typical pattern similar to that of an electrical discharge.

Photographs of the cavitation region formed by powerful ultrasonic radiators have also been provided [16, 17]. The diameters of the radiating surfaces of the radiators were greater than the sound wavelengths in the given liquid at the working frequencies. In these cases, plane acoustic waves are radiated into the liquid. The photographs show that at relatively low acoustic radiation intensity, the cavitation region is also located some distance away from the radiating surface, has an irregular pattern and is composed of thread-like collection of cavitation bubbles. As the radiation intensity goes up, however, the cavitation region approaches the radiating surface and grows in size. When the intensity reaches the value of, approximately, 1.5 W/cm^2, the cavitation region “sits” on the radiating surface and its shape starts to resemble an upside-down circular cone. The so-called “cone bubble structure” begins to form. Further radiation intensity increases have little effect on the shape and position of the cone bubble structure. The photographs in the abovementioned studies show that at high radiation intensity the cone bubble structure is in contact with the radiating surface. Reference [18] provides photographs of the radiating surface of a metal radiator which was utilized for a period of time to create high-intensity cavitation in a liquid. The surface of the radiator contains clear traces of metal degradation due to cavitation.

Therefore, it can be concluded with certainty that at high radiation intensities, acoustic cavitation starts at the surface of the acoustic radiator. This location in the liquid is known, according to theory, to have the lowest value of tensile strength due to the constant presence of adsorbed gas inclusions at the metal surface [4].

However, at low radiation intensities just above the cavitation threshold, the cavitation region is always formed at a significant distance away from the radiating surface, which contradicts the abovementioned theory. Clearly, the tensile strength of the liquid at any location away from the metal surface should be higher than near it, since the concentration of the preexisting bubbles (inceptions) that "weaken" the liquid at that location should diminish with time.

2.2. Justification for the Shock-Wave Approach

At low radiation intensity, harmonic acoustic wave is not capable of inducing cavitation even at the weakest location in the liquid near the radiating surface. Formation of cavitation away from the radiating surface in this case can be explained by the effect of the increase of the planar acoustic wave-front steepness during its propagation through a liquid. As a result of such an increase, at some location in the liquid a discontinuity in the wave profile is formed. Since such discontinuity is physically not possible in a continuous media, a shock-wave with a steep front is formed as a result. This effect has to do with the acoustic radiation-induced nonlinearity of the compressible media properties and is very well known and documented [19].

This explanation, however, seems contradictory to the common shock-wave theory, since the attainable amplitude of vibration velocity of the radiating surface is always much lower than the speed of sound in the pure liquid and, therefore, the necessary conditions for the creation of such a discontinuity in the wave profile are not fulfilled. The explanation may, nevertheless, still be valid due to the following two considerations. It is well known that during propagation of an acoustic wave of slightly lower intensity than the cavitation threshold, an ensemble of tiny bubbles is formed in the liquid. This occurs due to the so-called "rectified diffusion" [4]. It is also well known that the speed of sound in a liquid containing gas bubbles is significantly lower than that in a pure liquid [20, 21], and, under certain conditions, it may become similar to the amplitude of vibration velocity of the radiating surface.

It may, therefore, be considered that bubbles formed in an acoustic wave due to rectified diffusion help forming a discontinuity in the profile of the acoustic wave at a location away from the radiating surface by significantly lowering the sound speed in the liquid. Further, at the location of the discontinuity in the acoustic wave, these tiny bubbles begin to undergo such rapid nonlinear movements that they loose dynamic stability and, consequentially, rapidly multiply forming the cavitation region.

The abovementioned observations and analysis formed the basis of the shock-wave model of acoustic cavitation described in this section. The model shows how the primary energy of an acoustic radiator causing cavitation in a liquid is absorbed in the cavitation region owing to the formation of spherical shock waves inside each cavitation bubble. Calculation of the total energy absorbed in the cavitation region using the concept of a hypothetical spatial wave moving through the cavitation region is possible with this model using the classical system of the Rankine-Hugoniot equations. Additionally, the proposed model makes it possible to explain some newly discovered properties of acoustic cavitation of water that occur at extremely high oscillatory velocities of the radiating surfaces.

2.3. THEORY

Let us assume that an acoustic radiator emitting a plane-wave is used to generate cavitation in a liquid. The diameter of the radiator's output surface is comparable with the length of the acoustic wave in the liquid at the given frequency of vibrations. The vibration frequency is much lower than the resonance frequency of the cavitation bubbles. We assume that the liquid always contains an equilibrium concentration of dissolved gas as well as some cavitation nuclei (tiny spherical bubbles filled with the gas) and, consequentially, the liquid possesses no tensile strength during rarefaction caused by acoustic waves. As, for example, indicated in reference [4], water that has not been purified of gas inclusions ruptures at a negative acoustic pressure of, approximately, 1 bar. The density of the liquid with the tiny cavitation nuclei is taken to be equal to the density of the pure liquid, ρ_f. Surface tension of the liquid and the presence of stable (non-cavitational) gas bubbles are neglected. Thus, within the framework of the model, only the so-called low-frequency transient gas cavitation is considered. We, additionally, assume the liquid to be non-viscous, non-compressible and non-volatile.

Let us represent acoustic cavitation in the liquid as a sequence of the following events. When an acoustic rarefaction wave of certain amplitude passes through a volume of the liquid, an explosive growth of cavitation nuclei occurs, leading to the formation of the gas-filled cavitation bubbles. Possible parameters of such a rarefaction wave are described, for example, in [22]. A mixture of the spherical bubbles and the liquid is, therefore, formed. The gas dissolved in the volume of the liquid passes inside the free space formed by the bubbles. The density of the liquid medium, therefore, drops. At this point, the bubbles are so small, compared to the acoustic wavelength, that the liquid/bubble mixture can be considered a continuous medium. The rarefaction wave phase is followed by a compression wave phase, whose passage results in a collapse of all gas bubbles, restoring the density of the liquid to ρ_f. The reverse diffusion of the gas back into the liquid during compression is insignificant and should be ignored. This particular stage of acoustic cavitation completes the total cavitation cycle and is further considered here in great detail, since it is this stage that is mainly responsible for the sonochemical effects of acoustic cavitation.

2.3.1. Oscillations of a Single Gas Bubble

The problem of the liquid motion during compression of an empty spherical bubble in liquid was solved by Rayleigh (see reviews [4, 5]). On the basis of this solution and Ref. [19], the instantaneous pressure distribution in the liquid can be written as:

$$p = p_\infty + \rho_f \frac{\dot{U}r + 2U^2}{\xi} - \rho_f \frac{U^2}{2\xi^4} \qquad (1)$$

Here, p_∞ is the pressure in the liquid at infinity, U is the velocity of the bubble boundary (wall), $\xi = R/r$, r is the current bubble radius, and R is the current radial coordinate. For the boundary of a gas-filled bubble at $\xi = 1$, the following equality must be met:

$$p_g = p_\infty + \rho_f (\dot{U}r + \frac{3}{2}U^2) \qquad (2)$$

Here, p_g is the gas pressure in the bubble. This expression is the well-known Noltingk-Neppiras equation (see reviews [4, 5]).

For an empty bubble, taking $p_g = 0$ and $p_\infty = p_0$, integration of equation (2) gives Rayleigh's equations for the velocity of the bubble wall movement and the time of the bubble collapse:

$$U^2 = \frac{2p_0}{3\rho_f}\left(\frac{r_{in}^3}{r^3} - 1\right)$$

$$\tau = 0.915 r_{in}\left(\frac{\rho_f}{p_0}\right)^{0.5} \qquad (3)$$

Here, p_0 is the static pressure, and r_{in} is the initial bubble radius.

From equations (1) and (2), an expression for the instantaneous distribution of the pressure in liquid during the compression of a gas-filled bubble can be obtained:

$$p = p_\infty\left(1 - \frac{1}{\xi}\right) + \frac{p_g}{\xi} + \frac{\rho_f U^2}{2}\left(\frac{1}{\xi} - \frac{1}{\xi^4}\right) \qquad (4)$$

Let us single out a spherical liquid volume that includes a gas bubble. The gas bubble/surrounding liquid system has a certain acoustic compressibility, which determines the velocity of the propagation of small perturbations or the velocity of sound in this volume. Using the linearized form of the Noltingk-Neppiras equation, one can obtain an expression for the velocity of sound in such a system, as it was done, for example, in the work [21]. The velocity of sound, with the abovementioned assumptions taken into account, is determined using the following expression:

$$c = \left(\frac{p_g}{\rho_f \alpha (1 - \alpha)}\right)^{0.5} \qquad (5)$$

Here, α is the volumetric gas concentration in the singled-out liquid volume that includes a gas bubble. From equation (5) it can be seen that the velocity of sound at a given gas pressure in the bubble has a minimum at $\alpha = 0.5$. For example, at $p_g = 1$ bar the minimum velocity of sound $c_{min} = 20$ m/s. It should also be noted that the velocity of sound in the range $0.4 < \alpha < 0.6$ changes little.

A gas bubble is formed during the half-period of the liquid rarefaction in the acoustic wave. Under the abovementioned assumptions, this occurs at the moment when the pressure in the liquid near the wall of a cavitation nucleus decreases to zero, i.e. the negative acoustic pressure is equal to p_0. At that point, the gas pressure in the formed bubble is also very small. Further, during the subsequent period of increase in the acoustic pressure, the bubble is compressed, and the gas pressure in it also increases. During the subsequent compression half-period, in the singled-out liquid volume near the gas bubble wall a spherical flow in the direction of the bubble center is formed, which is described by equation (4). From equation (5) it is seen that the velocity of sound for the singled-out system gas bubble/surrounding liquid depends on the gas pressure in the bubble p_g and the value of coordinate ξ, along which the boundary of the singled-out volume passes. If we start reducing the singled-out volume, while the radius of the bubble and the gas pressure in it are constant, the velocity of sound in this system will fall to a certain limit and then will grow again. This means that in the considered spherical volume near the moving wall of the bubble, there is a critical spherical region, where the sound velocity, $c_{min,}$ is at the minimum at a given gas pressure in the bubble, p_g. The position of this region is determined from the condition $0.4 < \alpha < 0.6$. It is located close to the bubble wall in the coordinate range $1.18 < \xi < 1.35$. For the simplicity of further analysis of equation (4), it is taken that the velocity of the flow of the liquid particles in the critical region is equal to the velocity of the bubble wall movement, U.

In the model being considered, it is assumed that when the gas bubble/surrounding liquid system is compressed by the external pressure, p_∞, the velocity of the flow of the liquid particles in the critical region near the bubble wall increases to such a degree that at a certain gas pressure in the bubble, p_g, it reaches the minimum velocity of sound in the system under consideration, i.e. $U = c_{min.}$

At a ratio of the initial radius of an empty bubble to its current radius, $r_{in}/r = 2$, and static pressure, $p_0 = 1$ bar, the value of $U \approx 21$ m/s reached according to equation (3) is indeed close to $c_{min} = 20$ m/s.

Let us represent the pressure at infinity as a sum of the static and the acoustic (excessive) pressures, $p_\infty = p_0 + p'_\infty$ and transform equation (4) taking into account that $U = c_{min}$:

$$p = (p_0 + p'_\infty)(1 - \frac{1}{\xi}) + \frac{p_g}{\xi} + 2p_g(\frac{1}{\xi} - \frac{1}{\xi^4}) . \tag{6}$$

This expression describes the extreme condition of equilibrium of the system. Equation (6) shows that during compression of the flowing liquid, in the vicinity of the gas bubble a pressure impulse is formed, which is stationary with respect to the bubble wall. The amplitude of the excess pressure in this impulse is $p - p_0 = 1.4p_g + 0.5\ \delta p'_\infty$, where $\delta p'_\infty = (p'_\infty - p_0)$. This value is reached at the coordinate $\xi \approx 2$ located upstream from the critical region. As we show below, the quantity, $\delta p'_\infty$, does not need to be considered for small oscillation velocities of acoustic radiators.

When the velocity of the bubble wall motion exceeds the minimum velocity of sound, $U > c_{min}$, the equilibrium state described by equation (6) becomes destroyed, and the pressure in the liquid at the bubble wall downstream from the critical region decreases to p_0. The velocity of the bubble wall movement also reduces because the driving pressure difference decreases. At the same moment, the excessive pressure amplitude in the impulse increases stepwise up to the value $p - p_0 = 1.4p_0 + 0.5\ \delta p'_\infty$, since the boundary condition in equation (2) is changed and the pressure near the bubble wall becomes $p_g = p_0$. This occurs because the bubble pressure signal does not penetrate upstream from the bubble wall when $U > c_{min}$.

Due to destruction of the dynamic equilibrium (retardation of a part of the flow), the pressure impulse located in the liquid upstream from the critical section disintegrates and begins to move relative to the bubble boundary in the form of a converging spherical wave. The supposed instantaneous distribution of excessive pressure in the impulse near the gas bubble wall at $U = c_{min}$ is shown in Figure 2.

Phenomena similar in essence are observed during the breakup of arbitrary pressure discontinuity in a gas, during hydraulic impact, and during the flow of gases and gas-liquid mixtures through nozzles. See, for example, the works [6, 8], as well as the studies on Laval nozzles and water hammers.

In accordance with the assumed form of pressure distribution in a converging spherical wave shown in Figure 2, the excessive pressure at the bubble wall first increases smoothly up to the value of $p - p_0 = 1.4p_g + 0.5\delta p'_\infty$, and, accordingly, the gas pressure inside the bubble increases smoothly (isothermally) as well. Then, when an abrupt excess pressure jump (up to the value of $p - p_0 = 1.4p_0 + 0.5\delta p'_\infty$) approaches the bubble wall, a spherical shock wave is formed in the gas inside the bubble. The pressure jump itself, evidently, is equal to $1.4(p_0 - p_g)$. After focusing in the center of the gas bubble, the spherical shock wave is reflected, and the bubble "explodes" from the inside, breaking up into small fragments. The collapse of the gas bubble or, more precisely, its shock destruction occurs. Gas pressure and temperature inside the bubble during the focusing and the subsequent reflection of the shock wave reach very large, albeit theoretically restricted, values [19]. When the collapse of the gas bubble is completed, its small fragments are left in the singled-out liquid volume, which are equal in size to the original cavitation nuclei, and the density of the singled-out liquid volume becomes close to the initial liquid density, ρ_f. As we show below, when the oscillation velocities of the ultrasonic radiators reach very high values, cavitation may follow a different mechanism, which does not involve breaking the gas bubbles up into small fragments, but rather exhibits bubble behavior approaching that of an empty Rayleigh cavity.

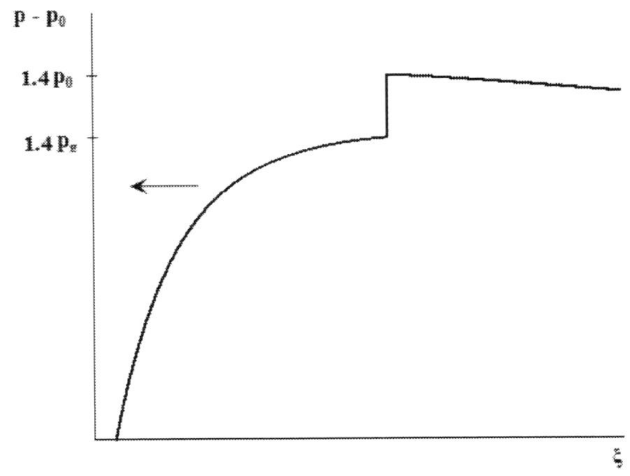

Figure 2. Instantaneous distribution of the excessive pressure in liquid near the cavitation bubble wall at $U > c_{min}$ is shown. The quantity $\delta p'_\infty$ is not taken into account.

This approach permits easily eliminating a seemingly clear contradiction that follows from the Noltingk-Neppiras equation: how can a gas-filled bubble implode with a very high rate if the gas pressure inside the bubble during compression rapidly increases, while the rate of the gas diffusion from the bubble, according to [4, 5], is negligible. In the proposed model, the gas bubble does not implode in the literal sense of the word, but is destroyed by a spherical shock wave reflected after focusing in its center. The presence of a well-known phenomena accompanying acoustic cavitation, such as sonoluminescence, erosion and dispersion of solids, emulsification of liquids, etcetera, can be well explained from this point of view. Additionally, the mechanism of the dissipation of the primary acoustic energy during the liquid cavitation becomes clear. This is the mechanism of the heating of a compressible medium in a shock wave, which is well described in the literature (see, for example, [19]).

2.3.2. Cavitation Region

During the rarefaction of a liquid in an acoustic wave, a mixture of a great number of spherical gas bubbles with the liquid (cavitation region) is formed. Let us call this gas-liquid mixture present in the cavitation region, the "continuum". In the previous section, the course of events during the collapse of a single bubble in some small volume of liquid was described. To extend these events over the entire continuum, a transition to spatial description is necessary. At that, the results of this transition must depend neither on the dimensions and the form of the continuum itself nor on the sizes and the spatial distribution of the bubbles in it.

During the compression stage, an acoustic radiator creates a pressure impulse in the liquid beyond the continuum in the form of a plane acoustic wave. Since the velocity of sound in the continuum is finite, the collapse of a multitude of gas bubbles located arbitrarily in the continuum must also occur simultaneously only in some narrow layer, as the impulse of the acoustic pressure approaches it, i.e. it must have a wave character. In the current model representation, the result of the superposition of many spherical shock waves, which are formed near each gas bubble during its collapse in a narrow layer of the continuum, is a spatial wave (SW) moving through the continuum. Such a representation is the most exact and visual way of extending the events occurring during a single gas bubble collapse, over the entire continuum.

In the real situation, the cavitation region in a liquid may take very complex, branched shapes. The spatial distribution of bubbles in the region also may be quite non-uniform and the sizes of the bubbles may vary. When the transition to the presented spatial description of cavitation is made, for the results to be independent of the shape of the cavitation region as well as of the spatial distribution and the sizes of the bubbles, in our initial equations we will further utilize hypothetical physical parameters related to the cavitation region as a whole. In other words, instead of operating with local values of density, changes in the internal energy and so on, we will use the values averaged over the whole cavitation region. As demonstrated below, these values disappear when further modifications of the fundamental equations are made.

The experimental investigations of acoustic cavitation described below conducted for the verification of the presented model were carried out using calorimetry of the entire environment and, therefore, provide only the spatially averaged values due to a relatively high thermal conductivity of the liquid. Therefore, the final purpose of the calculations following this model is the determination of a cumulative value of the changes in the internal energy of the environment, as a result of acoustic cavitation.

The spatial wave (SW) described above has a bore wave-like character, however, the continuum density and pressure inside the SW front change stepwise. This occurs because the cavitation bubbles collapse inside its front, following the process outlined in section 2.3.1. The presence of such a wave is the final stage of acoustic cavitation, within one cycle of the continuum rarefaction - compression. In other words, according to the model, it is assumed that the collapse of the gas bubbles occurs inside a relatively narrow front of a hypothetical SW, being formed and moving through the continuum in each compression half-period of an acoustic radiator.

The width of the SW front, inside which the collapse of the bubbles and the change of the continuum density occur, can be estimated as the product of the empty bubble collapse time, according to equation (3) and the wave front movement velocity with respect to the continuum, $h = c\tau$. A rough estimate for the wave front movement velocity can be made using expression (5). Then, at $\alpha = 0.1$ (taken from the literature data [22] and characteristic for the initial stage of acoustic cavitation) we obtain $h \approx 3r_{in}$. According to the estimation performed in the work [4], the maximum radius of a gas bubble in water does not exceed $2 \cdot 10^{-4}$ m, since larger bubbles rapidly rise to the surface. Hence, the value is: $h \leq 6 \cdot 10^{-4}$ m, which is smaller than the dimensions of the continuum itself by many orders of magnitude. Thus, the specified wave has a front that is very narrow relative to the dimensions of the entire continuum. Getting over this barrier, therefore, the physical parameters of the continuum change stepwise.

It is necessary, further, to establish a relation between the continuum parameters ahead of and behind the SW front, as well as the relationship between these parameters and the oscillatory velocity of an acoustic radiator. It is important to note that the velocity of the specified wave can be lower than the velocity of sound in the continuum.

The SW moving through the continuum is not only a physical abstraction used for the construction of the model, but can, apparently, exists in reality. In this case, however, we are not faced with an ordinary shock wave, which arises in a compressible continuum when the piston movement velocity is higher than the sound velocity in the continuum. Such shock waves in a gas-liquid suspension obtained by bubbling a gas through a liquid are described in detail in literature [21]. Here, it is assumed that in a gas-liquid suspension formed as a result of the liquid rarefaction in an acoustic wave, another type of bore wave-like shock waves may exist, which is associated with the radial movement of the liquid in the vicinity of each bubble.

It is well known that when a jump (discontinuity) of a physical quantity arises in a compressible continuum, a solution should be sought using the general conservation laws in the form of the Rankine-Hugoniot equations [19]. These equations reflect the ratios of the steady-state physical parameters of the compressible continuum before and after the passage of the shock wave front. Additionally, there appears a possibility to analytically calculate the values of important parameters, without considering in detail the transient processes inside the SW front, which are connected with the complex kinetics of a collapsing gas bubble.

Let us introduce the following designations: p_h is the pressure in the liquid phase of the continuum near the bubble wall after the SW passage; p_l, $\rho_l = \rho_f (1- \alpha_l)$, α_l are, respectively, the pressure in the liquid phase of the continuum near the bubble wall, the density and the volumetric gas content of the continuum before the SW passage. A scheme of the continuum flow is presented in Figure 3. It is assumed that a SW moves through the continuum, and that the gas bubbles collapse inside the narrow front of this wave. Also shown in this figure is the supposed pressure profile in the continuum.

Figure 4 shows the supposed processes occurring in one cycle of the acoustic cavitation of liquid. The pressure in the liquid phase of the continuum near the gas bubble wall in an arbitrary state is plotted on the ordinate, and the continuum specific volume is plotted on the abscissa. Line 1 represents the rarefaction of the continuum with cavitation nuclei in an acoustic wave. Line 2 represents a nonlinear process of the growth of cavitation bubbles in the rarefaction wave. Line 3 represents a preliminary compression of the continuum in an acoustic wave (for a single gas bubble, this corresponds to a rise in the gas pressure in the bubble on the smooth section of a converging spherical wave, as described in section 2.3.1). Line 4 represents the continuum's transition from one state to the other when the SW passes (for a single gas bubble, this corresponds to a rise in the gas pressure in the bubble on the steep section of a converging spherical wave, as described in section 2.3.1). In this scheme, it is assumed in advance that the velocity of the SW movement through the continuum can be lower than the sound velocity in the continuum itself ahead of SW. Additionally, the SW front itself serves as a source of the acoustic wave, propagating forward in the direction of the shock wave movement. In this connection, there is a preliminary compression of the continuum, and line 4 begins above the abscissa axis.

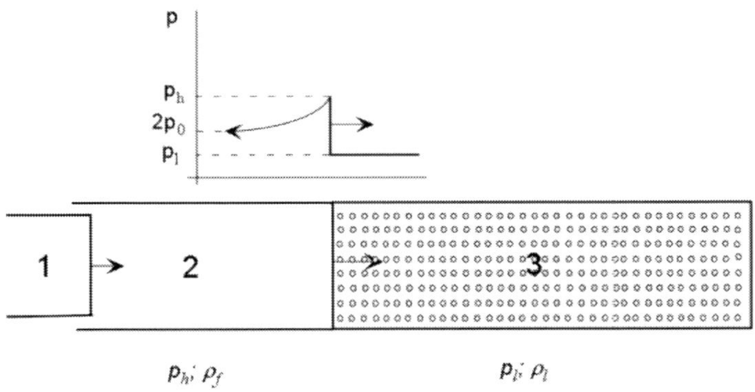

Figure 3. Schematic of the continuum's flow during compression is shown (1 – acoustic radiator, 2 – flow region after the SW passage, 3 – flow region before the SW passage).

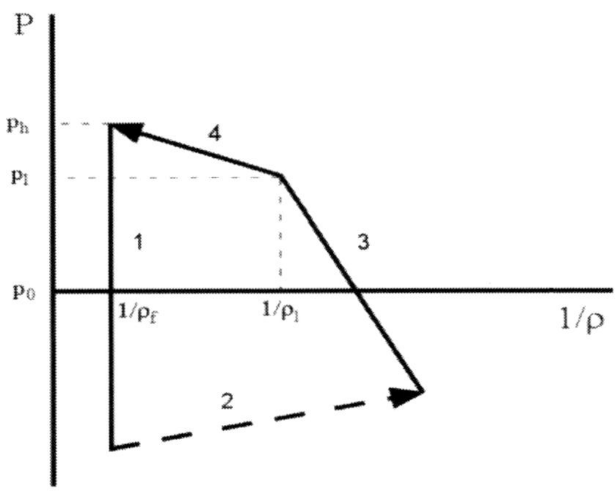

Figure 4. Processes occurring during acoustic cavitation are illustrated. Line 1 represents the rarefaction of the continuum with cavitation nuclei in an acoustic wave, line 2 represents a nonlinear process of the growth of cavitation bubbles in the rarefaction wave, line 3 represents a preliminary compression of the continuum in an acoustic precursor wave, line 4 represents the continuum transition from one state to the other when the SW passes.

This kind of an acoustic wave is called a precursor. The precursor does not cause the collapse and disintegration of the bubbles because of a small value of its amplitude. Similar representations are used for initially loose or porous environment. In such environment, during the compression phase, the shock-wave front is formed only due to the parameters of the compression process itself since this environment tends to change the specific volume of pores (cavities) abruptly (stepwise) under pressure [23-25].

Let us introduce the following additional designations: $p_l = p_0 + p'_l$, $p_h = p_0 + p'_h$; p'_l and p'_h are the excessive pressures in the liquid phase of the continuum near the bubble wall before and after the SW passage, respectively; u_l and u_h are the continuum flow velocities relative to SW before and after its passage, respectively; e_l and e_h are the specific internal energy of the continuum before and after the SW passage, respectively; v is the current oscillatory velocity of an acoustic radiator; v_t is the critical oscillatory velocity of an acoustic radiator, which corresponds to the cavitation onset (cavitation threshold). Note that a stepwise increase in the continuum density from ρ_l to ρ_f at the SW front corresponds to a change in pressure from p_l to p_h. The relative movement of the liquid and the gas bubbles is neglected.

Let us now write the system of conservation equations (Rankine-Hugoniot equations) for the continuum parameters on both sides of the SW front:

$$\begin{aligned} &\rho_l u_l = \rho_f u_h , \\ &p'_l + \rho_l u_l^2 = p'_h + \rho_f u_h^2 , \\ &\frac{p_0 + p'_l}{\rho_l} + \frac{u_l^2}{2} + e_l = \frac{p_0 + p'_h}{\rho_f} + \frac{u_h^2}{2} + e_h , \\ &v - v_t = u_l - u_h \end{aligned} \tag{7}$$

The fourth equation of system (7) shows that a change in the continuum's movement velocity getting over the SW front is equal to the excessive oscillatory velocity of an acoustic radiator, which exceeds the critical value, v_t.

This system of equations can be transformed to the following form:

$$I = \frac{(2p_0 + p'_l + p'_h)}{2}(v - v_t) , \tag{8}$$

$$\eta_l = \frac{(v - v_t)^2}{p'_h - p'_l}$$

Here, $I = (e_h - e_l)\rho_l u_h$ is the flux density of the energy dissipated inside the SW as a consequence of the dissipation processes related to the bubble collapse and $\eta_l = \alpha_l/\rho_l$ is the volume of all cavitation bubbles per unit mass of the liquid phase of the continuum before the SW passage.

The average flux density of the acoustic energy (acoustic energy intensity) absorbed in one acoustic wave period can be presented in the following way:

$$I_a = \frac{\omega}{2\pi} \int_0^{\pi/\omega} |I\, sin(\omega t)| dt = I/\pi \tag{9}$$

2.4. Set-up of Equations for Experimental Verification

For the resulting equations (8) to be verified experimentally, it is necessary to determine the particular values of p'_h, p'_l, η_l and v_t.

2.4.1. Low Oscillatory Velocities of Acoustic Radiator

From equation (6) and the analysis given in section 2.3.1, it follows that the maximum excessive pressure at the SW front is equal to $p'_h = 1.4p_0 + \delta p'_\infty$. As mentioned above, the liquid utilized for the construction of the theoretical model, does not possesses tensile strength during rarefaction. Consequentially, the explosive growth of the cavitation nuclei and their conversion into gas bubbles in the rarefaction wave takes place at the negative pressure equal to the static pressure, $p'_\infty = p_0$. It is possible to assume that for small oscillation velocities of the acoustic radiator near the cavitation threshold a symmetry of acoustic pressure amplitudes during the half periods of compression and rarefaction is conserved. Consequentially, in this case, $\delta p'_\infty = 0$ and $p'_h = 1.4p_0$. It will be shown below that for large radiator oscillatory velocities it is no longer possible to ignore the quantity $\delta p'_\infty$. Note that the value of $p'_h \approx 1.4p_0$ actually corresponds to the threshold of water cavitation, at least, in its initial stage. This fact was experimentally established in [26].

Above, it was assumed that during the rarefaction of a liquid in an acoustic wave, all gas dissolved in a unit volume of the liquid passes into the bubbles formed in this volume. The oscillations of the gas bubbles before the onset of their collapse are isothermal, and the mass of the gas in them does not change. From the analysis of equation (6) given in section 2.3.1, it follows that $p'_l = 1.4p_g$, hence, the condition $p_0\eta_0 = 0.71p'_l\eta_l$ must be met. Here, η_0 is the equilibrium volume of gas dissolved in a unit mass of the liquid at the pressure, p_0.

The quantity v_t is the critical oscillatory velocity of an acoustic radiator, which corresponds to the cavitation threshold. In view of the conditions described above, one can assume that for a plane acoustic wave, $(v_t)_{rms} = 0.71p'_\infty/\rho_f c_f = 0.71p_0/\rho_f c_f$.

It should be borne in mind that the value of v_t in each particular experimental case can be different from the specified theoretical value. This is connected with the fact that the practical value of v_t depends on a large number of different parameters of liquid (physical nature, purity degree, gas content, volatility, sample preparation history, etc.). Besides, v_t also depends on the conditions of the conducted measurements (frequency of ultrasound, degree of isolation from external radiation, temperature, etc.)

From the second equation of system (8) we obtain:

$$p'_l = \frac{1.4p_0^2\eta_0}{\eta_0 p_0 + 1.42(v - v_t)_{rms}^2} \tag{10}$$

Now from the first equation of system (8) in view of equations (9, 10) we obtain the final equation for the average flux density of the acoustic energy (intensity of acoustic energy) absorbed in the cavitation region:

$$I_a = 0.76p_0\left[1 + \frac{0.41p_0\eta_0}{\eta_0 p_0 + 1.42(v - v_t)_{rms}^2}\right](v - v_t)_{rms} \tag{11}$$

For the initial stage of acoustic cavitation, at a small value of $(v\text{-}v_t)_{rms}$, the final equation is as follows:

$$\frac{I_a}{p_0} = 1.07(v - v_t)_{rms} \tag{12}$$

It is important to point out that in equations (11, 12) the quantities related to the spatial distribution of gas bubbles in the continuum and their size, as well as the form and shape of the continuum itself are not present.

2.4.2. High Oscillatory Velocities of Acoustic Radiator

From the main system of equations (7), one can obtain the expression for the SW velocity relative to the unperturbed continuum, $u_l = \left[(p'_h - p'_l)/\rho_f \alpha(1-\alpha)\right]^{0.5}$. The ratio of u_l to the sound velocity, c, in the continuum according to equation (5), using equation (10) and taking into account that $p_g = 0.71p'_l$, can be written as:

$$\frac{u_l}{c} = \left(\frac{p'_h - p'_l}{p_g}\right)^{0.5} = \left(\frac{2(v - v_t)^2_{rms}}{p_0 \eta_0}\right)^{0.5} \tag{13}$$

From this expression, it is seen that at $(v\text{-}v_t)_{rms} \geq 1$ m/s, the SW movement must become supersonic, making it a real shock wave in the classical sense. When the SW movement is supersonic, a precursor is absent because it is absorbed by the faster shock wave. The density and the pressure of the gas inside the bubbles in this case are initially small since they are not compressed beforehand by the precursor. From the analysis of equation (10), it is seen that at $(v\text{-}v_t)_{rms} > 3$ m/s the gas pressure in such bubbles becomes approximately an order of magnitude lower than the static pressure, p_0, and continues to decrease. A spherical shock wave in rarefied gas inside such a bubble is not formed and, accordingly, the bubble does not break up into small fragments as a result of the collapse. The behavior of the bubble becomes close to the behavior of an empty Rayleigh cavity.

It is also important to keep in mind that the minimum width of the shock wave front in a gas is on the order of the molecule free path [19]. At a normal density of the gas, this distance is about 10^{-7} m. With a decreasing gas density, this distance increases proportionally and becomes close to the characteristic size of the bubble itself 10^{-5} m. Under these conditions, a spherical shock wave inside the bubble cannot be formed, and the bubble is compressed like a Rayleigh cavity.

At the final stage of the collapse of the bubble, the gas pressure in it increases to such a degree that it can hold back the liquid's pressure. At that, the pressure and temperature of the compressed gas can reach very high values (theoretically unrestricted under the assumptions of this model [19]). In this case, at the excess pressure, $p'_h = 1.4p_0$, the continuum behind the SW is a gas-liquid suspension with some density $\rho_h = \rho_f(1- \alpha_h)$. If the conditions identified in the beginning of section 2.3, assumed for the construction of the model, are to be met, the continuum behind the front of SW is additionally compressed by the acoustic radiator until density ρ_f is reached. This corresponds to a pressure increase at the SW front up to the value of $p'_h = 1.4p_0 + \delta p'_\infty = 1.4p_0 + 0.5c_h^2\delta\rho = 1.4p_0 + 0.5c_h^2\rho_f\alpha_h$, where $\delta\rho = \rho_f - \rho_h = \rho_f \alpha_h$ is the additional increase in the continuum's density behind the SW front, necessary to reach the quantity ρ_f, and c_h is the speed of sound in the gas-liquid suspension with density ρ_h. For high oscillatory velocities of acoustic radiator similar to the sound speed in the continuum, $p'_h = 1.4p_0 + \rho_f\alpha_h v^2_{rms}$, since in this case it can be taken that $c^2 = 2v^2_{rms.}$

The value of v_t is neglected. Since $\delta p'_\infty$ should be taken into account only at high v and the second term of equation (11), which corresponds to the excessive pressure p'_l, is negligible, we leave it unchanged. Let us now write equation (11) in the final form in view of equation (9):

$$I_a = 0.76\,p_0\left[1 + \frac{0.41\eta_0 p_0}{\eta_0 p_0 + 1.42(v - v_t)^2_{rms}} + \frac{0.29\rho_f\alpha_h v^2_{rms}}{p_0}\right](v - v_t)_{rms} \quad (14)$$

2.4.3. Interpretation of Experimental Results of Work [26]

A large series of experiments aimed at studying acoustic cavitation of water at low oscillatory velocities of acoustic radiator is presented in the work [26]. Experiments were conducted in degassed water with the concentration of the dissolved air equal to 30% of the nominal concentration in the equilibrium state at the room temperature and the normal static pressure.

For the interpretation of these data, let us introduce the following designations: $\Sigma I_a = 0.5(p'_h)^2\gamma = p_0^2\gamma$ is the total intensity of the acoustic energy radiated into water; $I_{a0} = 0.5(p'_h)^2\gamma_f = p_0^2\gamma_f$ is the intensity of the acoustic energy propagating beyond the bounds of the cavitation region.

Here, γ is the specific acoustic radiation admittance of the continuum, $\gamma_f = 1/\rho_f c_f$. The difference of these intensities is the intensity of the acoustic energy absorbed in the cavitation region. Thus, when compared with the theoretical results of the given model, the experimental values of γ for each oscillatory velocity obtained in [26] were recalculated by the following expression:

$$\frac{I_a}{p_0} = (\gamma - \gamma_f)p_0 \tag{15}$$

In representing the data of the work [26], the values of $(v_t)_{rms}$ were determined directly from the experimental plots of this work at the point of characteristic inflection.

2.5. Experimental Setup

To measure the acoustic energy absorbed in a cavitating liquid at increased static pressure p_0, an acoustic calorimeter described in section 3.2.3 of this book was used. Static pressure in the calorimeter was produced with compressed nitrogen. Settled tap water at 20^0 C was used. The static pressure, p_0, varied in the range 1.0 – 5.0 bar; the water density, $\rho_f = 998$ kg/m^3 sound velocity in the water, $c_f = 1500$ m/s; the volume of air dissolved in unit mass of water, $\eta_0 = 2.2 \cdot 10^{-5}$ m^3/kg. Each experimental point shown on the plots was obtained as a mean value of 10 measurements.

2.6. Experimental Results

Experimental data for small oscillatory velocities of an acoustic radiator, v, and different static pressures, p_0, are shown in Figure 5. The values of v_t used in the treatment of these experimental data were calculated from the expression $(v_t)_{rms} = 0.707 p_0/\rho_f c_f$ for different static pressures. Also shown in this figure are the experimental data from [26] for ultrasound frequencies of 19 and 28 kHz, closest to the frequency 17.8 kHz used in the present work, which are interpreted by equation (15). The values of the cavitation threshold obtained from the corresponding plots of [26] for both frequencies $(v_t)_{rms} = 0.08$ m/s. Figure 5 also shows the theoretical lines calculated from equations

(11) and (12), which are represented by the solid and the dotted lines, respectively.

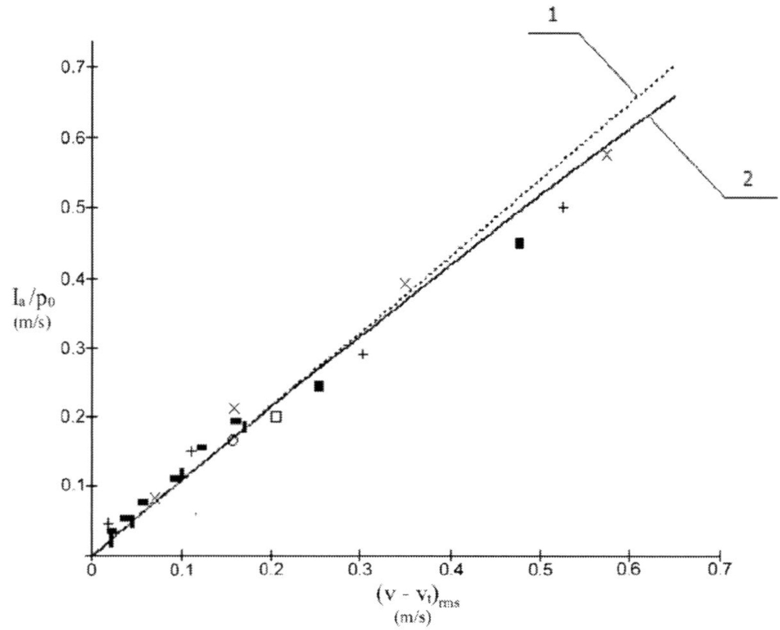

Figure 5. Intensity of acoustic energy absorbed in water at cavitation is shown as a function of the excessive oscillatory velocity of an acoustic radiator for pressures of × - 1 bar, + - 2 bar, ■ - 3 bar, □ - 4 bar, ○ - 5 bar, at frequencies of ▌- 28 kHz and ▬ - 19 kHz from the work [26]. Line 1 is plotted from equation (12); line 2 is plotted from equation (11).

A good agreement between the theoretical lines themselves and the experimental data with these lines at small values of v can be clearly seen. With increasing $(v\text{-}v_t)_{rms} > 0.2$ *m/s*, the experimental points diverge from the straight line plotted from equation (12) and approach the line plotted from equation (11).

Figure 6 shows the experimental results for all oscillatory velocities of the acoustic radiator, v, which were used in the experiments at normal static pressure, $p_0 = 1$ bar. Also shown in this figure are the theoretical lines plotted from equations (11) and (14). From Figure 6 it is seen that at intermediate values of v the experimental points are located near practically coincident lines plotted from equations (11) and (14), which are represented by the dotted and solid lines, respectively.

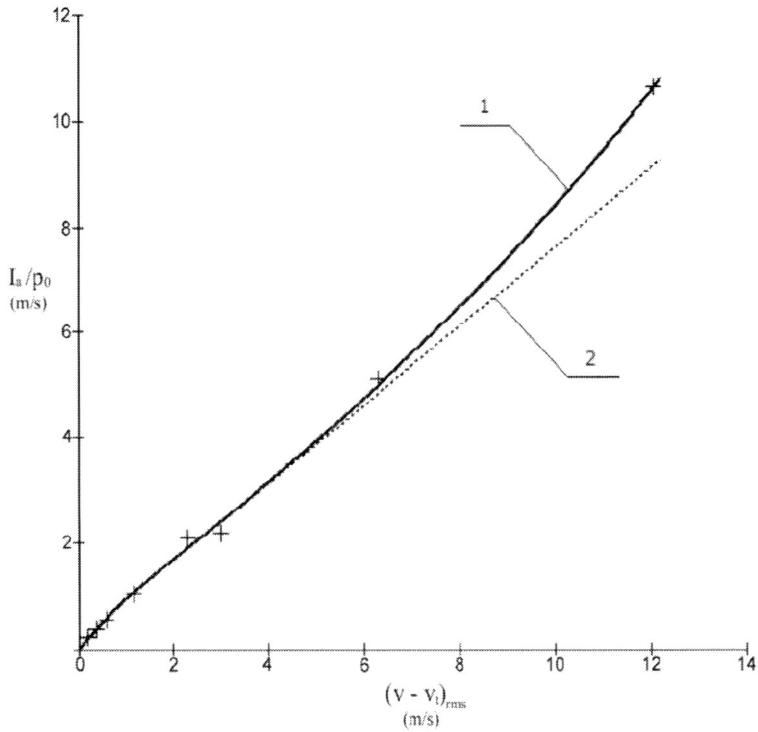

Figure 6. Intensity of acoustic energy absorbed in water at cavitation is shown as a function of the excessive oscillatory velocity of an acoustic radiator. Line 1 is plotted from equation (14); line 2 is plotted from equation (11).

At high oscillatory velocities, $(v\text{-}v_t)_{rms} > 3$ m/s, the specified theoretical relationships diverge, and the experimental points are located according to a more general relationship (14) at $\alpha_h = 0.4$. It can be seen that the theoretical and the experimental data are in good agreement up to the highest values of the oscillatory velocity, v.

A spread of the experimental points on the curve in Figure 6 in the region 2 m/s < $(v\text{-}v_t)_{rms}$ < 3 m/s is also observed. Here, the beginning of the divergence of the theoretical curves 1 and 2 is observed as well. These phenomena are, apparently, associated with the establishment of the supersonic regime of the SW movement and a considerable decrease in the gas pressure in the bubbles. The indication of the possibility of the supersonic regime of radiation at acoustic cavitation was first made in the work [27]. The phenomenon itself was called the second threshold of acoustic cavitation. The region located over the second threshold at $(v\text{-}v_t)_{rms}$

> 3 m/s was called the region of acoustic supercavitation. The closest related known phenomenon is called hydrodynamic supercavitation and is described, for example, in [28].

Since, as the stated theory assumes, at supercavitation the spherical shock wave is not formed in the gas inside the bubbles, at oscillatory velocities $(v-v_t)_{rms} > 3$ m/s the characteristic changes of the secondary effects of cavitation, which are used in the sonochemical technology, must be observed.

An experimental verification of this effect was conducted by observing the cavitation-induced ultrasonic dispersion of solid particles. During the experimental setup, it was assumed that the transition to the supercavitation regime should in some way be reflected in the manner in which the dispersion occurs. The experimental study was conducted during the ultrasonic dispersion of graphite particles with the initial size 200-250 μ in settled tap water under normal conditions. To avoid any possible influence of the reactor geometry on the results of the measurements, the acoustic calorimeter described in section 3.2.3 was used as an apparatus for dispersing. For the analysis of the relative transparency of the obtained dispersions, the degree of the light absorption (at the wavelength of 420 nm) in them was measured using a photo-colorimeter. From the measurement results presented in Figure 7 in relative units, it can be seen that the obtained curve reaches a maximum and then discontinues at 2.5 m/s $< (v-v_t)_{rms} < 3$ m/s. A subsequent smooth rise of this curve in the supercavitation region is also observed, which is most likely associated with the intense acoustic streaming, rather than with the effect of cavitation itself.

It appears that it is in the acoustic supercavitation region where the achievement of the highest possible temperatures during the compression of the rarefied gas inside the bubble oscillating as a Rayleigh cavity can be expected. Pressure at the bubble wall at the moment of focusing theoretically approaches infinitely high values because the gas compression is exerted by the moving dense bubble wall acting as a spherical plunger, rather than by a spherical acoustic wave [19]. In the same region, the highest intensities of the cavitation-induced sonochemical processes occurring at high temperatures may be observed. At the same time, processes connected with erosion, dispersion of solids and the like can be inhibited in the supercavitation region.

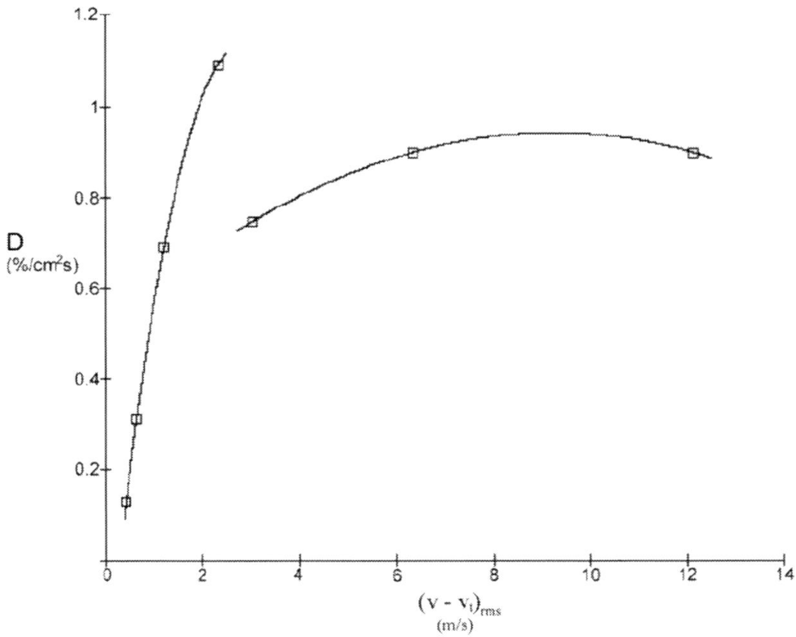

Figure 7. Dispersing effect of acoustic cavitation (dispersion of graphite powder in water) determined by the degree of the 420 nm wavelength light absorption is illustrated as a function of the excessive oscillatory velocity of an acoustic radiator.

2.7. Section Conclusions

The proposed shock-wave model of acoustic cavitation reflects real events occurring in water at cavitation since calculations based on the equations that follow from the model are in good agreement with the results of the experiments. The presented experimental data extend to the region of super-high oscillatory velocities of an acoustic radiator and agree well with the theoretical model. The model makes it possible to obtain the resulting equation for the calculations of the energy absorbed by liquids during cavitation without having to consider in detail all the complex processes of the absorption of the acoustic energy, which are connected with the nonlinear oscillations of the gas bubbles during their collapse.

Within the framework of this model, the existence of a transition from the subsonic regime of acoustic cavitation to the supersonic regime is predicted. The possibility of a change in the character of the oscillations of a cavitation bubble at high values of v is theoretically shown. The conducted experimental studies confirm such a possibility.

As will be shown below, simple algebraic expressions that follow from the proposed model can be used in practical engineering calculations for designing powerful ultrasonic horns for sonochemical reactors.

Chapter 3

Selection and Design of Main Components of High-Capacity Ultrasonic Systems

The greatest mechanical stress areas in a sonochemical reactor system are concentrated in the electromechanical transducer and the ultrasonic horn components. The same components are also exposed to the highest thermal loads, related to the formation and maintenance of acoustic waves. Selection of the appropriate electromechanical transducer type, therefore, is of great importance, as is the ultrasonic horn design and the choice of material from which it is constructed.

3.1. Electromechanical Transducer Selection Considerations

Ultrasonic transducers are devices used to convert electric energy coming from a power generator into mechanical energy in the form of ultrasonic vibrations. There are two main types of ultrasonic transducers used in the high-power ultrasonics field: magnetostrictive and piezoelectric (high-power piezoceramic).

For continuous flow-through liquid processing applications, magnetostrictive transducers have multiple advantages over the piezoelectric devices. These transducers are constructed from high-strength metallic alloys (5,000 – 7,000 MPa) and permit reaching high levels of acoustic power intensity (up to 100 – 150 W/cm^2). The main disadvantage of magnetostrictive transducers is their relatively low efficiency (below 50%). On the other hand, magnetostrictive transducers are electrically safe and do not overheat because they are relatively low voltage driven and liquid cooled. In addition, these transducers provide high total radiation powers and high output amplitudes, are very stable, reliable and do not age. These devices are, therefore, well suited for continuous long-term industrial operation under factory conditions and are ideal for industrial liquid processing with flow-through ultrasonic systems.

For comparison, the advantage of piezoelectric transducers is their high efficiency (up to 95%). These devices, however, are characterized by much lower levels of acoustic power intensity and relatively short life-spans due to low mechanical strengths of the involved materials (only about 15 – 30 MPa). Additionally, piezoelectric transducers are high-voltage driven and air cooled, which for some applications may make them an explosion hazard. They also can easily become overheated, which is why they cannot be used for extended periods of time or in high-temperature environments. These devices, however, are widely used in such important high-power ultrasonics fields as plastics welding, cleaning, machining, etc., where a pulsed-mode operation or lower amplitudes are appropriate. This explains these transducers' high popularity and availability. When used in liquid processing applications, however, piezoelectric transducers are frequently run at a much lower power than available, in a pulsed mode or with short periods of "on" time [29-31].

In view of the above discussion, we will only consider magnetostrictive transducers in this book.

3.2. High Power Acoustic Horn Design Principles

Despite being capable of producing much higher vibration amplitudes than piezoelectric devices, magnetostrictive transducers still cannot provide sufficient amplitudes for a correct operation of an ultrasonic reactor system. Acoustic (ultrasonic) rod horns are, therefore, used in conjunction with these transducers to amplify the vibration amplitude and deliver the ultrasonic energy to the working liquid. Commonly used acoustic horns (Figure 8), in general, consist of two cylindrical sections, input (larger diameter, in contact with transducer) and output (smaller diameter, in contact with the liquid), which are connected to each other by one transitional section, which may have a conical, exponential, catenoidal, or a more complex shape, or may be omitted all together (stepped horn) [32-34]. Although widely used, these horns suffer from an important limitation: they are incapable of providing matching between the transducers and the liquid loads, leading to an inefficient acoustic power transmission.

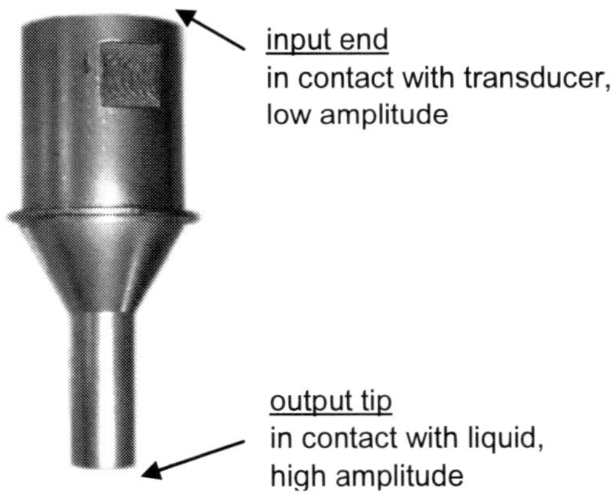

Figure 8. A typical high-gain converging horn is shown. High vibration amplitude of the output tip is achieved at the expense of the tip area.

For optimal operation, the maximum cross-sectional dimension of any portion of the resonant horn or transducer may not exceed, approximately, one quarter-wavelength of the corresponding longitudinal acoustic wave at the horn's resonance frequency [35]. Consequently, a common converging horn (for which the output diameter is smaller than the input diameter) with a maximal allowed input diameter always ends up having a working (output) tip diameter that is smaller than this limitation. The final size of the tip depends on the gain factor of the horn, and becomes reduced as the gain factor increases. This is problematic when the processes are carried out on an industrial scale, since the deposition of substantial acoustic power is needed to create acoustic cavitation in large volumes of water. While using converging horns permits increasing the acoustic energy intensity (or vibration amplitude) radiated by the transducer into the load quite effectively, it is impossible to achieve the technologically necessary levels of total radiated acoustic power, since the cross-sectional area of the horn tip in contact with the load is small. Therefore, it is intuitive that the use of converging horns does not permit transferring all available power from a transducer into a load.

To increase the total radiation area, the horns are sometimes connected to planar resonant systems, such as large discs or planes [36]. These additional elements, however, significantly complicate the construction of the horns, introduce additional mechanical connections and, therefore, reduce life span and reliability.

In this section we will describe design principles that have been successfully implemented in the development of a family of acoustic horns, whose shapes permit achieving high gain factors and large output surfaces simultaneously. These horns can be designed to accurately match an ultrasonic source (transducer) to a liquid load (water, in this case) at cavitation, maximizing the transference of the available acoustic energy into the load and creating a large cavitation zone. These devices are easy to machine and have well-isolated axial resonances and uniform output amplitudes.

3.2.1. Criteria for Matching a Magnetostrictive Transducer to Water at Cavitation

In an ideal case, without accounting for the internal losses, the highest acoustic energy intensity that a resonant magnetostrictive transducer can transmit into a load is limited by two main factors - the magnetostrictive stress saturation, τ_m (the maximum mechanical stress amplitude achievable due to the magnetostrictive effect for a given transducer material), and the maximum allowed amplitude of oscillatory velocity, limited by the fatigue strength of the transducer material, V_m, such that [37]:

$$\begin{aligned} \tau_m &= e_m E \phi_1 \\ V_m &= \sigma_m \phi_2 / \rho c \end{aligned} \tag{16}$$

where e_m is the deformation amplitude associated with τ_m, E is Young's modulus, ϕ_1 and ϕ_2 are the coefficients that take into account the features of the transducer construction [33, 37], σ_m is the stress amplitude of the material fatigue strength, ρ is the transducer material's density, and c is the thin-wire speed of sound in the material. The highest potential acoustic energy intensity radiated under conditions of perfect matching between the transducer and the load is represented by the quantity:

$$I_m = 0.5\tau_m V_m \tag{17}$$

It should be noted that the acoustic load under consideration, water at cavitation, has a purely active character, and, therefore, is appropriately described by the term "acoustic resistivity", r_a [26], such that $r_a = p_a / v$, where v is the amplitude of the output oscillatory velocity of acoustic horn and p_a is the acoustic pressure averaged over the entire radiating surface of the horn. Practically, this means that virtually all of the acoustic energy deposited into water at cavitation is converted into heat [38]. Under the term "matching" we will further mean supplying a magnetostrictive transducer with a multi-element acoustic horn having a gain factor, $G >> 1$ (G is defined as a ratio of the output to input oscillatory velocities, v / V_m, which allows the transference of a maximum of the available acoustic power of the transducer, I_m, into the load.

Acoustic energy intensity, I_a, generated in a purely active load by the longitudinal vibrations of an acoustic rod horn with an output oscillatory velocity amplitude, v, is represented by:

$$I_a = 0.5 r_a v^2 = 0.5 p_a v \tag{18}$$

Taking $I_m S_{in} = I_a S_{out}$ as a matching condition, we obtain:

$$\frac{\tau_m}{p_a} = GN^2 \tag{19}$$

where $N = \sqrt{S_{out} / S_{in}}$, S_{in} and S_{out} are, respectively, the input and the output cross-sections of the acoustic horn, while S_{in} is taken to be equal to the output cross-section of the magnetostrictive transducer, S_t (please see Figure 9). The left-hand side of equation (19) reflects the degree of under-loading of an acoustic transducer, and the right-hand side describes matching capabilities of an acoustic horn.

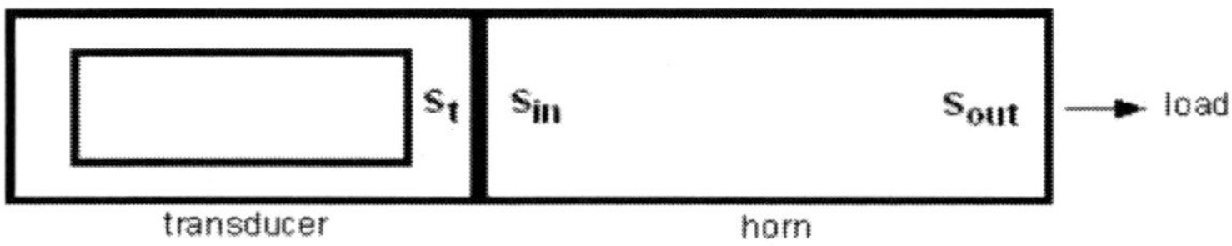

Figure 9. General schematic is shown, describing matching between an electromechanical transducer and a load achieved by using an acoustic rod horn of an arbitrary shape. S_{in} and S_{out} are, respectively, the input and the output cross-sections of the acoustic horn; S_t is the output cross-section of the electromechanical transducer.

As it was shown theoretically in section 2.4 and experimentally confirmed, the connection between the acoustic pressure, p_a, and the static pressure, p_0, during the well developed cavitation can be expressed by equations (11) and (14). To demonstrate this, let us consider the case of moderate (although much greater than the threshold value) amplitudes of ultrasonic vibration of the horn and apply equation (12). Assuming $v >> v_t$, and taking into account that $I_a = 0.5 p_a v$, we obtain for the amplitude value of p_a an expression $p_a \approx \sqrt{2} p_0$.

Therefore, the following can be written:

$$\frac{\tau_m}{p_a} = \frac{e_m E \phi_1}{\sqrt{2} p_0} \tag{20}$$

It is clear that for high vibration amplitudes, more complex expression based on equations (11) and (14) can be derived in a similar manner.

It is seen from equation (20) that the degree of under-loading of an acoustic transducer depends only on the characteristics of the transducer itself and the static pressure of water. Theoretically, for most common magnetostrictive materials, the calculated values of τ_m / p_a are between 15 and 44. In this calculation, the values of $p_0 = 10^5$ N/m^2 and $\phi_1 = 0.45$ were assumed. However, for a real magnetostrictive transducer, which is an electro-acoustic instrument, the maximum acoustic energy intensity generally does not exceed 70 -100 W/cm^2. This is due to such limitations as an insufficient ultrasonic generator power, voltage and current rating of the electrical wire forming the transducer's coil, cooling system capacity, etc. Consequentially, the practical values of the degree of under-loading are much lower than the corresponding theoretical limits for the magnetostrictive materials themselves, and for most models are between 5 and 10.

It is less evident how to use the right-hand side of equation (19), which reflects the matching capabilities of a horn. In this case, before the resonance calculation of a matching horn it is necessary to determine the maximum acoustic energy intensity for the utilized magnetostrictive transducer, $I_m = \eta W_e$. Then, from (17) and (19) we obtain:

$$GN^2 = \frac{\sqrt{2\eta W_e}}{p_0 V_m}, \tag{21}$$

were W_e - specific (with respect to S_{in}) electrical power of the magnetostrictive transducer, η - its efficiency (commonly $\eta = 0.5$). The next step should be selecting an optimal, from the technological standpoint, range of the values for the gain of the horn, G, which is commonly determined during the preceding laboratory studies of a given process. It is then easy to derive the value for N necessary for the resonance calculations of the matching horn and construction of the ultrasonic reactor.

In spite of a variety of types and shapes of the acoustic horns known from the literature and used in practice, until recently none existed for which the relationship $GN^2 > 1$, when $G > 1$, would hold true. It is, however, clear that in order to be able to match magnetostrictive transducers to water at cavitation, it is necessary to utilize acoustic horns that would meet the matching criterion, $GN^2 > 1$.

3.2.2. Five-Elements Matching Horns

3.2.2.1. Design Principles

The theory of acoustic horns is based on the mathematical problem of longitudinal vibrations in multi-element rods that have cylindrical elements as well as elements of variable cross-sections [39]. We will consider only the horns of axially symmetric shapes. Other types of horns (for example, wedge-shaped) can be considered in an analogous way. In the current work, we will restrict the discussion to the five-element horns, although no theoretical restriction for the number of elements exists.

We assume that during the passage of stress waves through a horn, the wave front remains planar, while the stresses are uniformly distributed over the horn's cross-section. This assumption limits us to "thin" horns, whose resonance lengths significantly exceed their diameters. For all practical purposes, it is sufficient to require that the maximum cross-sectional dimension of any portion of a resonant horn not exceed, approximately, one quarter-wavelength of the corresponding thin wire acoustic wave at the horn's resonance frequency [35].

The schematic and the designation of parameters for a general case of a five-element rod horn are given in Figure 10, where two possible situations are presented: a horn with $d_1/d_3 > 1$ is shown by the solid line; a horn with $d_1/d_3 < 1$ is shown by the dotted line. Under the assumed approximation, the problem is reduced to one-dimension, and it is limited to the consideration of elements with variable cross-section of only conical shape.

For a steady-state mode, the equation of vibrations for displacements, u, takes the following form:

$$u'' + \frac{1}{S} S'u' + k^2 u = 0. \qquad (22)$$

where $k = \omega / c$ is the wave number, $\omega = 2\pi f$ is the angular frequency of vibrations, and f is the frequency of vibration.

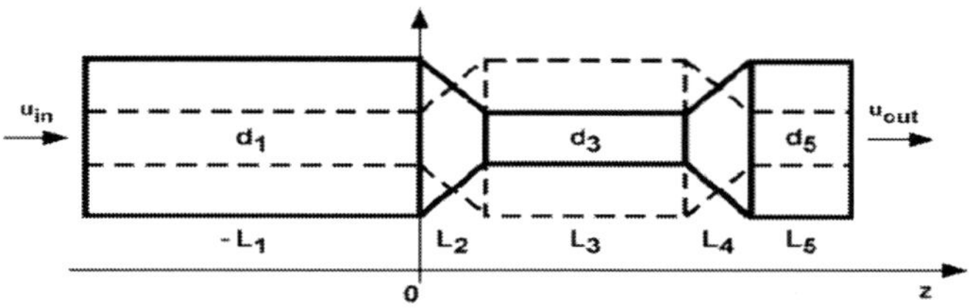

Figure 10. Schematic defining the parameters of a five-element matching horn is shown. The horn having $d_1 / d_3 > 1$ is shown by a solid line, and the horn with $d_1 / d_3 < 1$ is shown by a dotted line. Parameters $L_1 - L_5$ correspond to the lengths of each element.

The solutions of equation (22) for each of the horn's elements can be written as:

$$
\begin{aligned}
u_1 &= A_1 \cos kz + B_1 \sin kz && -L_1 < z < 0 \\
u_2 &= F(A_2 \cos kz + B_2 \sin kz) && 0 < z < L_2 \\
u_3 &= A_3 \cos kz + B_3 \sin kz && L_2 < z < L_2 + L_3 \\
u_4 &= F(A_4 \cos kz + B_4 \sin kz) && L_2 + L_3 < z < L_2 + L_3 + L_4 \\
u_5 &= A_5 \cos kz + B_5 \sin kz && L_2 + L_3 + L_4 < z < L_2 + L_3 + L_4 + L_5
\end{aligned}
\tag{23}
$$

Then, using the boundary conditions for the horn's element, we obtain the system of equations for displacements, u , and strains, u' .

At $z = -L_1$, $u_1 = u_{in}$, $ES_1 u_1' = -F_{in}$, $F_{in} = 0$

$$
\begin{aligned}
&A_1 \cos kL_1 - B_1 \sin kL_1 = u_{in}; \\
&EkS_1 (A_1 \sin kL_1 + B_1 \cos kL_1) = -F_{in}
\end{aligned}
$$

At $z = 0$, $u_2 = u_1$, $u_2' = u_1'$

$$
\begin{aligned}
&FA_2 = A_1; \\
&F'A_2 + FB_2 k = kB_1; \\
&\beta = (d_1 - d_3)/L_2 d_1; \\
&F = 2/d_1;\ F' = F\beta
\end{aligned}
$$

At $z = L_2$, $u_3 = u_2$, $u_3' = u'_2$

$$
\begin{aligned}
&A_3 \cos kL_2 + B_3 \sin kL_2 = F(A_2 \cos kL_2 + B_2 \sin kL_2); \\
&-kA_3 \sin kL_2 + kB_3 \cos kL_2 = (F'B_2 - FkA_2)\sin kL_2 + (F'A_2 + FkB_2)\cos kL_2; \\
&\beta = (d_1 - d_3)/L_2 d_1; \\
&F = 2/d_3;\ F' = -F\ (L_2 - 1/\beta);
\end{aligned}
\tag{24}
$$

At $z = L_2 + L_3$, $u_4 = u_3$, $u'_4 = u'_3$

$$F[A_4 \cos k(L_2 + L_3) + B_4 \sin k(L_2 + L_3)] = A_3 \cos k(L_2 + L_3) + B_3 \sin k(L_2 + L_3);$$
$$(F'B_4 - FkA_4) \sin k(L_2 + L_3) + (F'A_4 + FkB_4) \cos k(L_2 + L_3) = \\ = -kA_3 \sin k(L_2 + L_3) + kB_3 \cos k(L_2 + L_3);$$
$$\beta = (d_3 - d_5)/L_4 d_3;$$
$$F = 2/d_3;\ F' = F\beta$$

At $z = L_2 + L_3 + L_4$, $u_5 = u_4$, $u'_5 = u'_4$

$$A_5 \cos k(L_2 + L_3 + L_4) + B_5 \sin k(L_2 + L_3 + L_4) = \\ = F[A_4 \cos k(L_2 + L_3 + L_4) + B_4 \sin k(L_2 + L_3 + L_4)];$$
$$-kA_5 \sin k(L_2 + L_3 + L_4) + kB_5 \cos k(L_2 + L_3 + L_4) = \\ = (F'B_4 - FkA_4) \sin k(L_2 + L_3 + L_4) + (F'A_4 + FkB_4) \cos k(L_2 + L_3 + L_4);$$
$$\beta = (d_3 - d_5)/L_4 d_3;$$
$$F = 2/d_5;\ F' = -F/(L_4 - 1/\beta)$$

At $z = L_2 + L_3 + L_4 + L_5$, $u_5 = u_{out}$, $u'_5 = 0$

$$A_5 \cos k(L_2 + L_3 + L_4 + L_5) + B_5 \sin k(L_2 + L_3 + L_4 + L_5) = u_{out};$$
$$-A_5 \sin k(L_2 + L_3 + L_4 + L_5) + B_5 \cos k(L_2 + L_3 + L_4 + L_5) = 0$$

The gain factor of the horn can be expressed as:

$$G = \left|\frac{u_{out}}{u_{in}}\right| = \left|\frac{A_5 \cos k(L_2 + L_3 + L_4 + L_5) + B_5 \sin k(L_2 + L_3 + L_4 + L_5)}{A_1 \cos kL_1 - B_1 \sin kL_1}\right| \quad (25)$$

where $F = 2/d_n$, d_n is the diameter of the corresponding cylindrical element of the horn, A_n and B_n are the constant coefficients for the corresponding elements of the horn, L_n is the length of the corresponding element of the horn, n is the order number of the horn element, β is the cone index of the horn element with variable cross-section, u_{in} and u_{out} are the amplitudes of displacements at the horn input and output, respectively. The boundary conditions for the force acting on the horn's input, $F_{in} = 0$, and for the strain at the horn output, $u_5' = 0$, in this system of equations indicate that the horn has a total resonance length and does not have an acoustic load.

From the system of equations (24), one can obtain all necessary characteristics of a five-element horn: lengths and diameters of the elements, gain factor, distribution of vibration amplitudes, and distribution strains along the horn. From this system of equations, it is also easy to obtain solutions for any horns with conical elements (for example, with fewer than five elements). Horns with other shapes of the variable cross-section elements (for example, exponential or catenoidal) can be considered in an analogous way, taking into account the variation of sound velocity in such elements.

3.2.2.2. Analysis of Five-Element Horns

To solve the system of equations (24) and to present results in a convenient form, a computer program has been written that allows all the indicated above characteristics of five-element horns to be obtained in real time for subsequent analysis. The input parameters are: operating frequency of the horn, acoustic properties and fatigue strength of the horn's material, and the diameter-to-length ratios of the horn elements. For the convenience of comparison of horn parameters, we further assume $d_1/d_5 = 1$.

From all possible solutions of the system of equations (24), only the series of five-element acoustic horns will be considered, which will be referred to as "barbell horns". This series of horns, in the authors' opinion, is the most useful for industrial applications, in particular, for building industrial ultrasonic reactor systems.

Figure 11 shows a half-wave barbell horn and its design parameters. A photograph of this horn is also presented in Figure 14 (b). The maximum value of the matching capability of this horn is $GN^2 \approx 4$. The resonance length of this horn corresponds to one half of the ultrasonic wavelength in the metal from which the horn is constructed, with dispersion taken into account. Its small resonance dimensions are convenient in terms of manufacturing and minimizing the side surface radiation, and should be particularly noted. Some useful parameters of this type of horn are presented in Table 1.

Table 1.

kL_1	G	KL_5
0.5	1.79	0.215
1.0	3.17	0.128
1.5	3.78	0.093
2.0	3.46	0.058

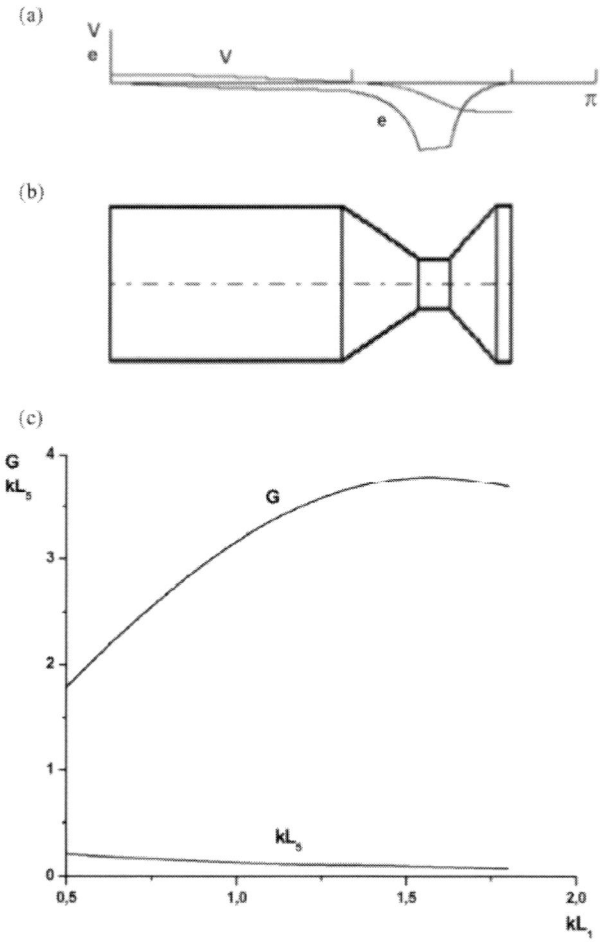

Figure 11. Half-wave barbell horn is shown with $d_1 = d_5$; $d_1/d_3 = 3.0$; $kL_2 = 0.5$; $kL_3 = 0.2$; $kL_4 = 0.3$, along with (a) the distribution of the oscillatory velocity, V, and strain, e, along the horn; (b) drawing of the horn; (c) plot of the distribution of the horn's parameters.

Figure 12 shows a spool-shaped barbell horn and its design parameters. This horn is atypical because its main radiating surface is lateral, and it mainly radiates a cylindrical wave into the load, as opposed to a plane wave radiated by other matching horns. Given a symmetric form of the horn, the gain factor is always equal to 1, the node of displacements is located in the middle, and lateral surfaces move in anti-phase. When using lateral radiation, the horn's matching capabilities are quite high since there are no limitations on the overall length. When such horns are connected into a sequential string (radiating part of the long spool-shaped barbell horn, shown in Figure 17 (a)), they can radiate a cylindrical wave of high total power into the load and produce a well-developed cavitation region of a large volume. Some useful parameters of this type of horn are presented in Table 2.

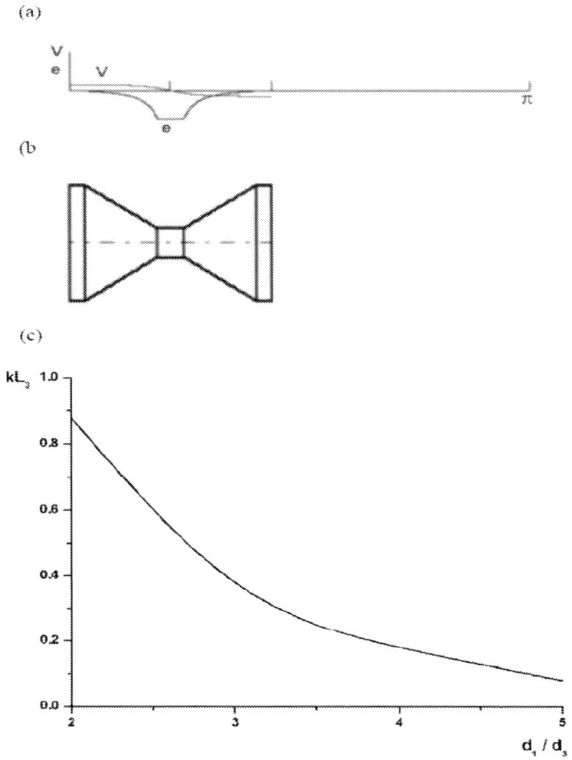

Figure 12. Symmetrical spool-shaped barbell horn is shown with $d_1 = d_5$; $kL_1 = kL_5 = 0.1$; $kL_3 = kL_4 = 0.5$, along with (a) the distribution of the oscillatory velocity, V, and strain, e, along the horn; (b) drawing of the horn; (c) plot of the distribution of the horn's parameters.

Table 2.

d_1/d_3	KL_3
2.0	0.877
3.0	0.384
4.0	0.179
5.0	0.085

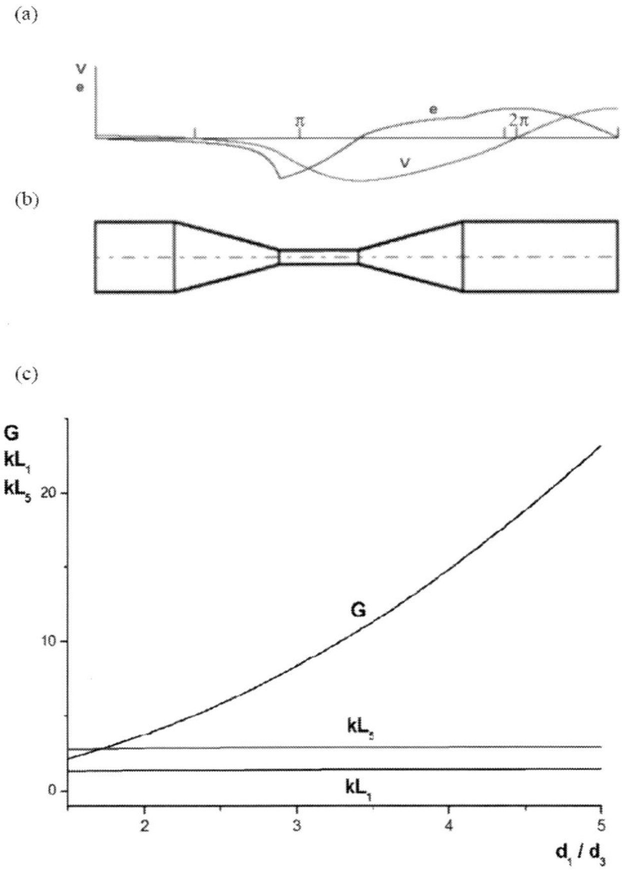

Figure 13. Full-wave barbell horn is shown with $d_1 = d_5$; $kL_1 = kL_3$; $kL_2 = kL_4 = 0.5$, along with (a) the distribution of the oscillatory velocity, *V*, and strain, *e,* along the horn; (b) drawing of the horn; (c) plot of the distribution of the horn's parameters.

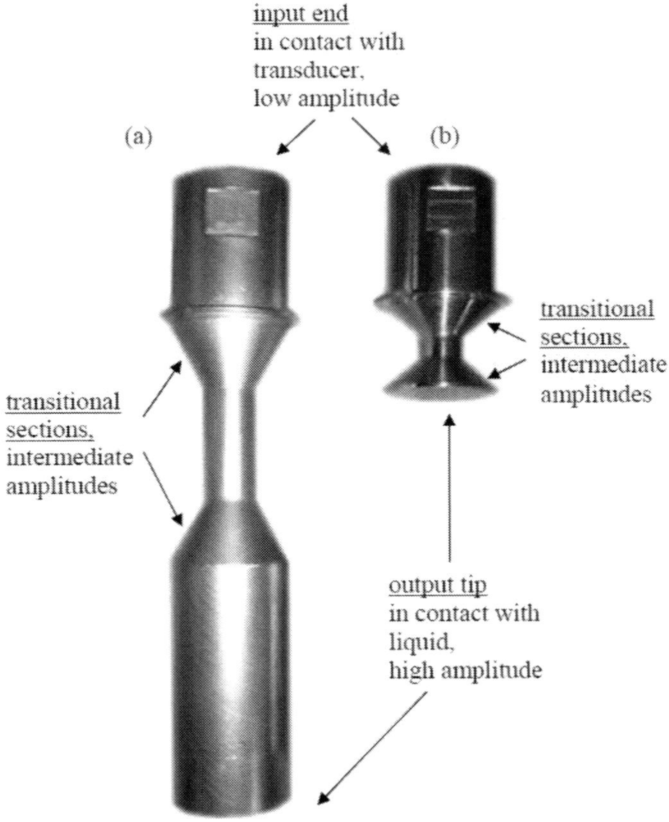

Figure 14. A full-wave (a) and a half-wave (b) high-gain barbell horns are shown. High vibration amplitude of the output tip is achieved without having to sacrifice the tip diameter. These particular barbell horns have output tip diameters of 65 mm and provide ultrasonic amplitudes (a) up to 120 microns peak-to-peak and (b) up to 80 microns peak-to-peak.

Above, we have considered the horns whose lengths were less than or close to half the length of the acoustic wave in the rod, the so-called half-wave barbell horns. The system of equations (24) also allows one to obtain solutions for full-wave barbell horns. One of such horns intended for the radiation of a plane acoustic wave of a very high power into water is a full-wave barbell horn shown in Figures 13 and 14 (a). Its design parameters, as a function of d_1/d_3, are presented in Figure 13 (c). The matching capabilities of the full-wave barbell horn can reach the values of

$GN^2 = 20$ or more. These horns are very promising for the matching of high-power magnetostrictive transducers that have large cross-sections.

For example, the highest design power radiated into the water at cavitation by this horn, made of high-quality titanium alloy, taking into account the fatigue strength limitations and limitations on output diameter under normal static pressure, is about 5 kW at a frequency of 20 kHz.

Due to the significant potential of the full-wave barbell horn for the industrial applications of ultrasound, we also provide its exact parameters in Table 3. These parameters are convenient for practical calculations.

Table 3.

d_1/d_3	G	kL_1	kL_2	kL_5
1.5	2.176	1.383	0.405	2.853
2.0	3.527	1.290	0.693	2.725
2.5	4.918	1.245	0.916	2.640
3.0	6.285	1.224	1.099	2.574
3.5	7.597	1.216	1.253	2.519
4.0	8.834	1.215	1.386	2.470
4.5	9.987	1.217	1.504	2.426
5.0	11.049	1.222	1.609	2.384

3.2.3. Experimental Results

For the experimental verification of the described horn design principles we have chosen the full-wave barbell horn of the type shown in Figures 13 and 14 (a). Direct calorimetric measurement of acoustic energy transmitted by this horn into water at cavitation was selected as a method of this horn's performance evaluation, as well as for obtaining experimental results presented in section 2.6. The measurements of the acoustic energy absorbed in the cavitation region were conducted with the apparatus shown in Figure 15. Settled tap water at a temperature of 20 ^{0}C was used. The apparatus was based on an acoustic radiator consisting of a titanium horn connected to a magnetostrictive transducer, which operated at the resonance frequency of 17.8 kHz. The working power of the ultrasonic generator coupled to the magnetostrictive transducer was 5 kW. The oscillation amplitude of the magnetostrictive transducer was kept constant in all experiments at 1.67 m/s (rms). It was measured by placing a magnetic ring with an inductive coil on the transducer next to its output surface. Voltage was created in the coil as the transducer oscillated. The amplitude of this voltage corresponded to the oscillation amplitude and was measured by an oscilloscope. Prior calibration of this device was performed, in which the vibration amplitude was measured directly by a microscope.

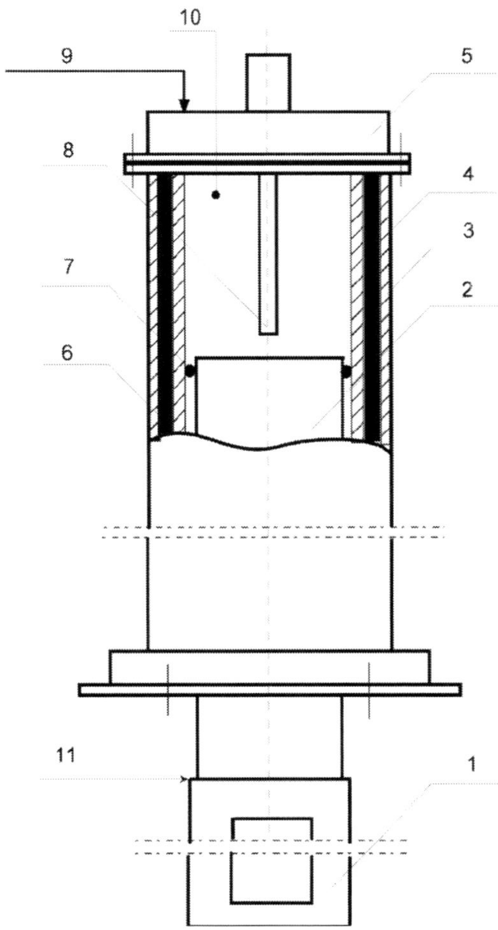

Figure 15. Schematic of acoustic calorimeter is presented. 1-magnetostrictive transducer, 2-replaceable full-wave barbell horn, 3-external wall of calorimeter, 4-heat insulation gasket, 5-cover with porous sound-absorber, 6-internal wall of calorimeter, 7-sealing ring, 8-set of thermocouples, 9-gas supply, 10 – microphone, 11-point of control over amplitude of transducer vibrations.

A set of replaceable full-wave barbell horns was constructed to provide the necessary stepped change in the amplitude of the oscillatory velocity of the output end in contact with water. The set consisted of nine such horns with different gain factors (greater or smaller than unity), all of which had equal input and output diameters of 60 mm. Maximum oscillation velocity of some of these horns reached very large values, close to maximum theoretically possible for the best titanium alloys. Greatest achieved oscillation velocity was 12 m/s (rms). Therefore, maximum gain factor for the set was 7.2.

Static pressure in the calorimeter was produced with compressed nitrogen. The measurements of the resulting temperature of water were performed using a set of thermocouples. A change in the temperature of water during ultrasonic treatment was not more than 2 – 5 ^{0}C.

Experimentally measures acoustic energy intensity levels absorbed in the cavitation area are presented above in Figures 5 and 6. The dispersing effect of acoustic cavitation is shown in Figure 7.

Performance verification of the horns with different gain factors conducted during the experiments showed that all of them possessed resonance and gain characteristics well corresponding to the theoretically predicted values. In no case was it necessary to make any adjustments to the horns after they were originally machined.

The region of the acoustic energy intensity with the values above 10^5 W/m^2 is very little studied, especially from the technological standpoint. The reason for this, from our perspective, is that the traditional cone-shaped horns, widely used in ultrasonic technology, are incapable of providing a large total radiation power, since their oscillation amplitudes are inversely proportional to the areas of their output surfaces. At large gain factors, the output surface area becomes very small, which complicates the development of sonochemical reactors capable of processing significant volumes of liquids. Thus, for example, a traditional stepped horn having an input diameter of 60 mm and a gain factor of 7.2 has the output diameter of, approximately, 20 mm. Therefore, at the maximum experimentally achieved acoustic energy intensity of 10^6 W/m^2, this stepped horn is capable of depositing no more than 300 W into its liquid load. Our full-wave barbell horn, used in the experiments presented in this section, on the other hand, delivers, approximately, 2.7 kW of total power, providing a power transfer efficiency increase by almost an order of magnitude.

3.3. Section Conclusions

Matching a magnetostrictive transducer to water is a matter of selecting a horn type that fulfills the expression (19) at a given gain factor G, and of subsequent calculation of its resonance dimensions with the use of the system of equations (24). The most powerful horn, among the designs described above, is the full-wave barbell horn, which was chosen for the experimental investigations. During the experiments, evaluation of a set of such horns with different gain factors showed that all of them had the resonance and the gain factor characteristics that corresponded very well to those predicted theoretically. It was also experimentally verified that matching of the acoustic horns with water at cavitation, according to the theory described above, is truly established for all values of the output oscillation velocities of the horns.

It should be noted that matching an acoustic transducer to a load using an acoustic horn is not the only possible method of matching. Another powerful matching factor, which results from the specific properties of water at cavitation, is the static pressure, p_0, according to the expression (11) and the experimental results. It is clear that the best results are obtained when these two matching techniques are used jointly.

In conclusion, we would like to add that barbell horns also perform well in non-aqueous liquids and solutions with significant viscosity, and permit building very effective ultrasonic reactors, suitable for treatment of such liquids, for example oils, epoxy resins, honey, polymer melts, metal melts, etc.

Photographs presented in Figure 16 illustrate the primary (a) and secondary (b) cavitation zones formed during the operation of a full-wave barbell horn having an output diameter of 65 mm providing acoustic energy intensity of $2x10^5$ W/m^2 in the primary cavitation zone below output tip.

In certain applications of powerful ultrasonic systems, however, it is more important to increase the residence time of the working liquid in the reactor, than to maximize the output amplitude. This is especially important during preliminary preparation for further high-amplitude processing, such as during pre-dispersion, pre-emulsification, treatment of high-viscosity liquids, etc. In these cases, it is convenient to utilize a long spool-shaped barbell horn, incorporated into a reactor chamber. Figure 17 shows such a horn (a) as well as the cavitation zones formed by it in a relatively viscous liquid, glycerin (b). This figure shows that two well developed secondary cavitation zones are formed near the two "necks" of the long spool-shaped barbell horn, constructed as two spool-shaped barbell horns connected in series.

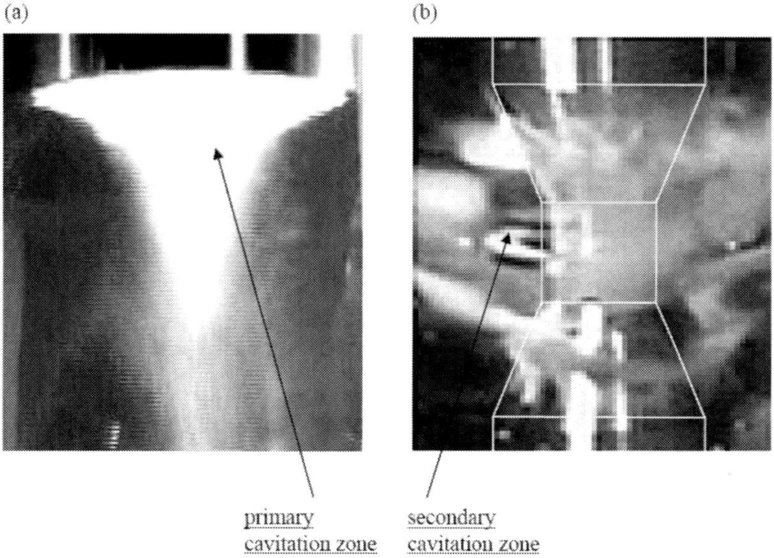

Figure 16. Experimentally obtained photographs of well developed stable cavitation zones are shown. The zones were created in an unrestricted volume of water by a barbell horn, having the following operational parameters: output tip diameter – 65 mm, ultrasound frequency – 18 kHz, acoustic energy intensity – 20 W/cm^2. Part (a) shows the primary cavitation zone under the horn tip; part (b) shows the secondary cavitation zone produced near the neck of the barbell horn (marked with a white line).

In semi-industrial ultrasonic reactor systems with relatively low transducer power (1 - 2 kW), it is convenient to use half-wave barbell horns, shown in Figure 14 (b). These horns are compact and have minimal losses due to the side-surface radiation.

All photographs shown above were obtained using ultrasonic equipment produced by Industrial Sonomechanics, LLC (ISM). Videos showing primary and secondary cavitation zones produced by barbell horns operating at a range of ultrasonic amplitudes are available at ISM's website [40].

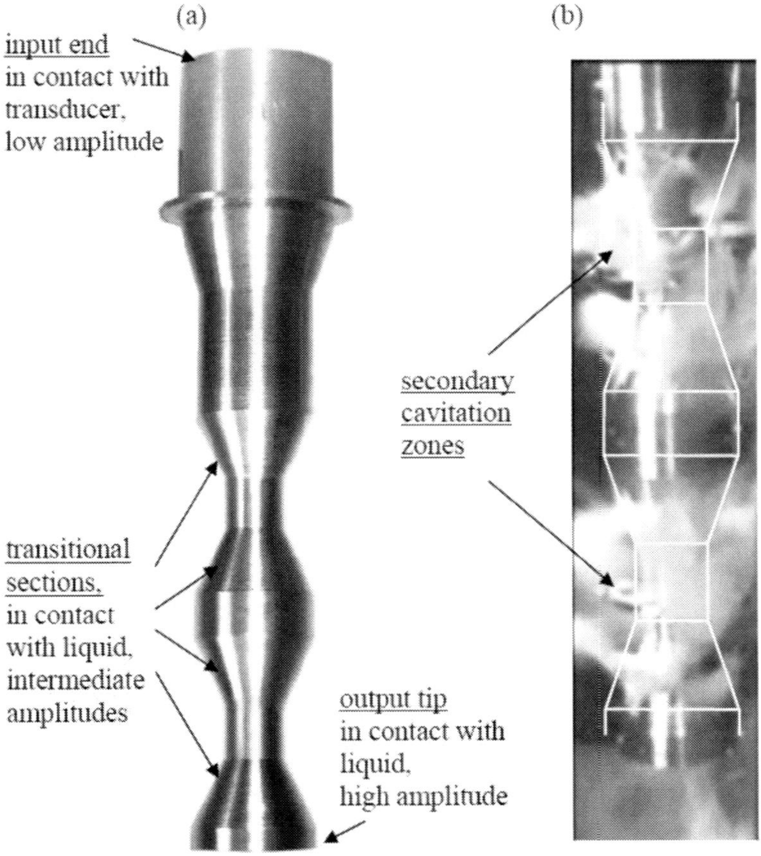

Figure 17. Photograph of a long spool-shaped barbell horn is shown in part (a). Photograph taken during operation of this horn in glycerin is displayed in part (b), showing multiple secondary cavitation zones formed near its transitional sections.

Chapter 4

Ultrasonic Reactor Chamber Geometry

During a flow-through ultrasonic process, it is important to make sure that all working liquid is directed through the active cavitation zone, otherwise inhomogeneous processing may result, leading to a lower-quality product. Eliminating the low cavitation intensity areas in the reactor also helps increase the power density that the system can deposit into a liquid load. Optimization of the ultrasonic reactor chamber geometry, therefore, leads to an improvement in the technological effects obtained during the operation of the reactor.

In a common, unoptimized reactor chamber the treated liquid enters through the inlet at the bottom, passes through the primary cavitation zone of a horn, Figure 16 (a), flows along the horn's side surface and comes out through the outlet at the side of the chamber at the top. If a barbell horn is utilized, there is also a secondary cavitation zone near the transitional sections, as shown in Figure 16 (b), which accounts for approximately 20 % of the total radiated ultrasonic power. An optimized reactor chamber design would efficiently direct all treated liquid through both of these cavitation zones.

It has been explained above that the shape of a well developed cavitation zone formed at the bottom of a barbell horn resembles an upside-down circular cone. Therefore, it is beneficial to shape the bottom of the reactor chamber in the same manner, as shown in Figure 18. An approximately 20 % increase in the absorbed acoustic energy can be achieved due to the presence of a cone insert at the bottom of the reactor chamber, which optimizes the volume and the shape of the main cavitation zone at the output tip of the barbell horn [2]. To take the full advantage of the secondary cavitations zone, a liquid deflector ring may be inserted near the neck of the barbell horn (its second cylindrical section), as shown in Figure 18. Supplying the reactor chamber with both of these features dramatically improves the homogeneity of ultrasonic exposure of the working liquid and increases the total power deposition.

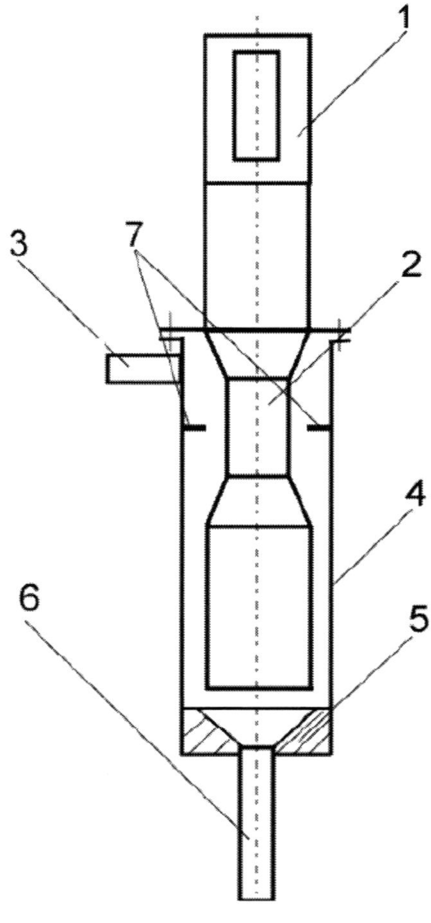

Figure 18. Schematic of an optimized flow-through ultrasonic reactor is presented, where 1 – electro-acoustical transducer, 2 – barbell horn, 3 – working liquid outlet, 4 – reactor chamber, 5 – upside-down circular cone insert, 6 – working liquid inlet, 7 – circular reflection surface.

Chapter 5

FINAL REMARKS

Industrial implementation of ultrasonic reactor systems has not reached its full potential. This is especially true when processes require high ultrasonic amplitudes, for example in production of nanoemulsions or nanodispersions. On the other hand, a large number of laboratory studies exist that demonstrate high potential effectiveness of ultrasonic processing in these and other areas [41, 42].

Since prior to the introduction of barbell horns the ultrasonic amplitude amplification was commonly done with converging horns, high-amplitude industrial-scale ultrasonic equipment was not available. Consequentially, transferring the results of many laboratory studies involving high-amplitude ultrasound to the plant floor has not been possible. Low-amplitude (below 30 μ_{pp}) industrial ultrasonic equipment has been around for several decades. This equipment, however, has had limited capability to translate optimized ultrasonic processes to commercial scale due to its inability to provide high-intensity cavitation in large reactor volumes. Additionally, this equipment has generally relied on piezoelectric transducer designs, which for industrial-scale liquid processing applications suffer from several important limitations compared with magnetostrictive devices.

The ultrasonic cavitation theory and main hardware design principles presented in this book provide the background necessary to construct high-capacity industrial ultrasonic systems with up to 10,000 L/h processing capability, able to operate at extremely high ultrasonic amplitudes in excess of 150 μ_{pp}. Using these systems, any laboratory study results can be directly implemented on industrial scale by simply increasing the horn tip diameter and the corresponding reactor volume and boosting the power of the generator and the transducer. All of the process parameters optimized during the laboratory study (ultrasonic amplitude, reactor residence time, pressure, etc.) can be retained, while the production rate may be increased by orders of magnitude.

REFERENCES

[1] S.L. Peshkovskiy, M.L. Friedman, and W.A. Hawkins, *Ultrasonic Rod Waveguide-Radiator*. 2004, Industrial Sonomechanics, LLC: U.S. Patent #7,156,201.

[2] S.L. Peshkovsky and A.S. Peshkovsky, *High Capacity Ultrasonic Reactor System*. 2008, Industrial Sonomechanics, LLC: International Application #PCT/US08/68697.

[3] T.G. Leighton, *Bubble population phenomena in acoustic cavitation.* Ultrason. Sonochem., 1995. 2: p. 123.

[4] H.G. Flynn, *Physics of acoustic cavitation in liquids*, in *Physical Acoustics, principles and methods*, W.P. Mason, Editor. 1964, Academic Press: New York and London. p. 78-172.

[5] M.S. Plesset and A. Prosperitty, Ann. Rev. Fluid Mech., 1977. 9.

[6] J.L. Laborde, A. Hita, J.P. Caltagirone, and A. Gerard, *Fluid dynamics phenomena induced by power ultrasounds.* Ultrasonics, 2000. 38: p. 297-300.

[7] W. Lauterborn and C.D. Ohl, *Cavitation bubble dynamics.* Ultrason. Sonochem., 1997. 4(2): p. 65-75.

[8] G. Servant, J.L. Laborde, A. Hita, J.P. Caltagirone, and A. Gérard, Spatio-temporal dynamics of cavitation bubble clouds in a low frequency reactor: comparison between theoretical and experimental results. Ultrason. Sonochem., 2001 8(3): p. 163-74.

[9] J. Klíma, A. Frias-Ferrer, J. González-García, J. Ludvík, V. Sáez, and J. Iniesta, Optimisation of 20 kHz sonoreactor geometry on the basis of numerical simulation of local ultrasonic intensity and qualitative comparison with experimental results. Ultrason. Sonochem., 2007 14(1): p. 19-28.

[10] W. Lauterborn, T. Kurz, R. Geisler, D. Schanz, and O. Lindau, *Acoustic cavitation, bubble dynamics and sonoluminescence.* Ultrason. Sonochem., 2007 14(4): p. 484-91.

[11] R. Mettin, P. Koch, W. Lauterborn, and D. Krefting. Modeling acoustic cavitation with bubble redistribution. in Sixth International Symposium on Cavitation, CAV2006. September 2006. Wageningen, The Netherlands.

[12] G. Servant, J.L. Laborde, A. Hita, J.P. Caltagirone, and A. Gérard, *On the interaction between ultrasound waves and bubble clouds in mono- and dual-frequency sonoreactors.* Ultrason. Sonochem., 2003 10(6): p. 347-55.

[13] R. Mettin, S. Luther, C.D. Ohl, and W. Lauterborn, *Acoustic cavitation structures and simulations by a particle model.* Ultrason. Sonochem., 1999 6(1-2): p. 25-9.

[14] S. Luther, R. Mettin, P. Koch, and W. Lauterborn, *Observation of acoustic cavitation bubbles at 2250 frames per second.* Ultrason. Sonochem., 2001 8(3): p. 159-62.

[15] J. Appel, P. Koch, R. Mettin, D. Krefting, and W. Lauterborn, *Stereoscopic high-speed recording of bubble filaments.* Ultrason. Sonochem., 2004 11(1): p. 39-42.

[16] A. Moussatov, C. Granger, and B. Dubus, *Cone-like bubble formation in ultrasonic cavitation field.* Ultrason. Sonochem., 2003. 10: p. 191–195.

[17] A. Moussatov, R. Mettin, C. Granger, T. Tervo, B. Dubus, and W. Lauterborn. *Evolution of acoustic cavitation structures near larger emitting surface.* in *World Congress on Ultrasonics, WCU2003.* September, 2003. Paris, France.

[18] P. Diodati and G. Giannini, *Cavitation damage on metallic plate surfaces oscillating at 20 kHz.* Ultrason. Sonochem., 2001 8(1): p. 49-53.

[19] Y.B. Zel'dovich and Y.P. Raizer, Physics of Shock Waves and High-Temperature Hydrodynamic Phenomena. 1966, New York: Acad. Press.

[20] R.A. Thuraisingham, *Sound speed in bubbly water at megahertz frequencies.* Ultrasonics, 1998. 36(6): p. 767-773.

[21] L. Van Vijngaarden, *Annual review of fluid mechanics*, in *Annual Review Inc.* 1972: Palo Alto. p. 369.

[22] L.D. Rosenberg, *High-intensity ultrasonic fields.* 1971, New York: Plenum Press.

[23] W. Herrmann, A Constitutive Equation for the Dynamic Compaction of Ductile Porous Materials. J. Appl. Phys., 1969. 40: p. 2490.

[24] A.D. Resnyansky and N.K. Bourne, *Shock-wave compression of a porous material.* J. Appl. Phys., 2004. 95: p. 1760-1769.

[25] M.G. Salvadori, R. Skalak, and P. Weidlinger, *Stress Waves in Dissipative Media.* Transactions New York Academy of Science, Ser. II, 1959. 21(5): p. 427-434.

[26] K. Fukushima, J. Saneyoshi, and Y. Kikuchi, *Ultrasonic Transducers*, ed. Y. Kikuchi. 1969, Tokyo: Corona Publ. Co.

[27] N.B. Brandt, A.D. Yakovlev, and S.L. Peshkovsky, Russ. Tech. Phys. Let., 1975. 1(10): p. 460.

[28] R.T. Knapp, J.W. Daily, and F.G. Hammitt, *Cavitation.* 1970 New York: McGraw-Hill.

[29] P. Chand, C.V. Reddy, J.G. Verkade, and D. Grewell. *Enhancing Biodiesel Production from Soybean Oil using Ultrasonics.* in *ASABE Paper No. 8*. 2008. St. Joseph, MI, USA.

[30] A.K. Singh, S.D. Fernando, and R. Hernandez, *Base-catalyzed fast transesterification of soybean oil using ultrasonication.* Energy & Fuels, 2007. 21: p. 1161-1164.

[31] G. Towerton, *The use of ultrasonic reactors in a small scale continuous biodiesel process.* 2007, G&M Global Enterprises Inc.: Amarillo, TX, USA. p. 1-4.

[32] U.S. Bhirud, P.R. Gogate, A.M. Wilhelm, and A.B. Pandit, Ultrasonic bath with longitudinal vibrations: a novel configuration for efficient wastewater treatment. Ultrason. Sonochem., 2004. 11: p. 143-147.

[33] E. Eisner, *Physical Acoustics*, in *Methods and Devices, Part B*, W.P. Mason, Editor. 1964, Acad. Press: New York.

[34] S. Sherrit, S.A. Askins, M. Gradziol, B.P. Dolgin, X.B.Z. Chang, and Y. Bar-Cohen, *Novel Horn Designs for Ultrasonic/Sonic Cleaning, Welding, Soldering, Cutting and Drilling.* Proceedings of the SPIE Smart Structures Conference, San Diego, CA, 2002. 4701: p. Paper No. 34.

[35] J.W. Rayleigh (Strutt), *The Theory of Sound.* 1945, New York: Dover Publications.

[36] J.A. Gallego Juárez, G. Rodríguez Corral, E. Riera Franco de Sarabia, C. Campos Pozuelo, F. Vázquez Martínez, and V.M. Acosta Aparicio, *A Macrosonic System for Industrial Processing.* Ultrasonics, 2000. 38: p. 331-336.

[37] Y. Kikuchi, *Ultrasonic Transducers*, ed. Y. Kikuchi. 1969, Tokyo Corona Publ. Co.

[38] E.A. Neppiras, Measurements in liquids at medium and high ultrasonic intensities. Ultrasonics, 1965. 3(1): p. 9-17.

[39] L.G. Merkulov and A.B. Kharitinov, *Theory and analysis of sectional concentrators.* Sov. Phys. - Acoust., 1959(5): p. 183-90.

[40] http://www.sonomechanics.com

[41] J.P. Canselier, H. Delmas, A.M. Wilhelm, and B. Abismaïl, *Ultrasound Emulsification—An Overview.* Journal of Dispersion Science and Technology, 2002. 23(1): p. 333 – 349.

[42] T.J. Mason and J.P. Lorimer, Applied Sonochemistry: Uses of Power Ultrasound in Chemistry and Processing. 2002, Weinheim: Wiley-VCH. 303.

INDEX